T0305495

Remanufacturing and Remanufacturability Assessment for the Circular Economy

This book presents decision support tools that can be used in the early design stage to analyze the feasibility of a product and its components for remanufacturing. It also covers how to design a product specifically for remanufacturing and offers supporting case studies.

This is a comprehensive solutions guide for remanufacturing decision-making. The book illustrates an approach that can be used at the product End-of-Life (EOL) stage to generate optimized recovery plans for the returned products. Opportunities for Industry 4.0 to support remanufacturing along with case studies are included to showcase the decision-making tools.

Remanufacturing and Remanufacturability Assessment for the Circular Economy: A Solutions Guide will be of interest to practitioners, business professionals and researchers that work in the industrial and manufacturing sectors. Those involved with supply chain management and advanced technologies associated with Industry 4.0, sustainability and integrated techniques of circular supply chains will also find this book very useful.

Remanufacturing and Remanufacturability Assessment for the Circular Economy
A Solutions Guide

Yang Shanshan, S. K. Ong, and A. Y. C. Nee

CRC Press
Taylor & Francis Group
Boca Raton London New York

CRC Press is an imprint of the
Taylor & Francis Group, an **informa** business

First edition published 2023
by CRC Press
6000 Broken Sound Parkway NW, Suite 300, Boca Raton, FL 33487-2742

and by CRC Press
4 Park Square, Milton Park, Abingdon, Oxon, OX14 4RN

CRC Press is an imprint of Taylor & Francis Group, LLC

© 2023 Yang Shanshan, S. K. Ong, and A. Y. C. Nee

ISBN: 978-1-032-23085-6 (hbk)
ISBN: 978-1-032-23088-7 (pbk)
ISBN: 978-1-003-27559-6 (ebk)
ISBN: 978-1-032-38101-5 (eBook+)

DOI: 10.1201/9781003275596

Typeset in Times
by codeMantra

Contents

Preface

Remanufacturing is a key enabler for sustainable production due to its effectiveness in closing the loop on material flow, extending product life cycle and reducing production waste and emission. It is the process of returning an End-of-Life (EOL) product to 'as-new' condition, through the processes of disassembling, cleaning, inspecting, reconditioning, replacing and reassembling the components of a part or product, and thus it provides a material recirculation loop within the product system. This solutions guide presents decision support tools, to be used at the early design stage, to analyze the feasibility of a product and its components for remanufacturing, and to facilitate the design of the product for remanufacturing. At the same time, this solutions guide proposes a comprehensive approach, to be used at product EOL stage, to generate optimized recovery plans for the returned products.

The decision to include remanufacturing as a part of a product life cycle should be made early at the design stage, as about 80% of the cost of the product is determined at this stage (David and Anderson, 2014). However, this decision involves rather complex considerations, as business, engineering, market, economic and environmental factors can all affect the success of a remanufacturing endeavor. Besides, the uncertainties involved in various dimensions, such as quality, quantity and timing of product return, have further complicated the remanufacturing strategy planning issue. In this regard, this book presents a logical way of determining whether certain products or components are feasible for remanufacturing at the initial product design stage, through evaluating and analyzing a comprehensive list of decision-making factors. A genetic algorithm is adopted to determine a Pareto set of optimal EOL strategies, which will facilitate the effort of decision makers to maximize the environmental benefits of remanufacturing for a given economic profit. This book also presents a proactive approach to improve the remanufacturability of the products/components, namely, design for remanufacturing. As pointed out in various studies, the barriers to the remanufacturing process can largely be traced back to the initial product design stage, and this has highlighted the importance of early integration of the design characteristics that can enhance the product or process remanufacturing efficiency. This book aims to steer a product design toward remanufacturability from four major design aspects, namely, material selection, material joining methods, structural design and surface coating during the product design stage. The generated design alternatives are further evaluated and compared from both the remanufacturing and life cycle perspectives, through a multi-criteria decision-making technique and life cycle assessment (LCA) method, respectively, to improve the effectiveness and robustness of the design decisions.

Besides the early design stage, this book presents methods to examine the suitability of products and their components for remanufacture or disposal during the EOL return stage based on their return conditions. The framework includes both qualitative and quantitative analyses to address the operational and technological considerations for product remanufacturing and optimize the environmental and economic performance. Probability theory is utilized in the framework to analyze the impact of the quality of the returned products on EOL decision making. To represent the product structure hierarchy and the interconnections among the components of a product, the Hierarchical Attributed Liaison Graph (HALG) is used, allowing both complete and partial disassembly strategies to be considered during EOL strategy planning.

Finally, this book discusses the ways in which Industry 4.0 revolution can help effectively address these issues and unlock the potential of remanufacturing for the circular economy.

Authors

Dr. Yang Shanshan is the deputy group manager of the Virtual Manufacturing Group at the Advanced Remanufacturing and Technology Centre (ARTC), at A*STAR – Agency for Science, Technology, and Research, Singapore. Her research focuses on data-driven sustainability, strategy planning for product End-of-Life (EOL) disposition, reverse supply chain, and life cycle assessment and design. She is currently leading and working on platform solutions for connecting manufacturing, logistic, customer and End-of-Life data for improved supply chain optimization, sustainability evaluation and decarbonization. Dr. Yang received her Ph.D. from the National University of Singapore (NUS) where she studied product design for remanufacturing and remanufacturability assessment. During her Ph.D. study, she also spent one year at the Golisano Institute for Sustainability, Rochester Institute of Technology (RIT) working on research projects on material selection for remanufacturing and End-of-Life strategy planning. Dr. Shanshan has published over 16 journal and conference papers, with 634 citation.

S. K. Ong is an associate professor in the Mechanical Engineering Department at the National University of Singapore. She is a fellow of CIRP, The International Academy for Production Engineering. Her research interests are virtual and augmented reality applications in manufacturing, ubiquitous manufacturing, assistive technology and rehabilitation engineering. She has published 8 books and papers in over 320 international refereed journals. She received the 2004 Eugene Merchant Outstanding Young Manufacturing Engineer Award from the US Society of Manufacturing Engineers. She has also received various other awards including the 2009 Emerging Leaders Award in Academia by the US Society for Women Engineers.

A. Y. C. Nee is currently a Professor Emeritus in the Department of Mechanical Engineering, National University of Singapore. He received his Ph.D. and D.Eng. from the University of Manchester in 1973 and 2002, respectively. He is a Fellow of CIRP, SME and the Academy of Engineering Singapore. He was the president of CIRP in 2012 and Gold Medal Recipient of SME in 2014. His awards include IEEE Kayamori Award (1999), Norman A Dudley Award, International Journal of Production Research (2003) and Joseph Whitworth Prize, the Institution of Mechanical Engineers (2009). Dr. Nee was elected as one of the Asia's top 100 scientists by *Asian Scientist Magazine* in 2016. His research interests include tool, die, and fixture design, augmented reality applications, digital twin and remanufacturing. He has published over 500 papers and 20 books with 20,200 citations.

List of Abbreviations

AHP	Analytic Hierarchy Process
CC	Cumulated Cost
CED	Cumulated Energy Demand
CGI	Compacted Graphite Iron
DfA	Design for Assembly
DfE	Design for Environment
DfRem	Design for Remanufacturing
DfX	Design for X
DRRA	Design for Remanufacturing and Remanufacturability Assessment
EHS	Environmental Health and Safety
ELECTRE	Elimination and Choice Expressing the Reality
EOL	End-of-Life
HALG	Hierarchical Attributed Liaison Graph
IR	Independent Remanufacturers
LCA	Life Cycle Assessment
MCDM	Multi-Criteria Decision Making
MA	Manufacturing
MEP	Material Extraction and Processing
NSGA	Non-Dominated Sorting Generic Algorithm
OEM	Original Equipment Manufacturers
PROMETHEE	Preference Ranking Organization Method for Enrichment Evaluation
PSS	Product Service Systems
PTB	Product Take-Back
QFD	Quality Function Deployment
RE	Remanufacturing
RFID	Radio Frequency Identification
RoHS	Restriction of Hazardous Substances
SAMSG	Sustainable Automotive Materials Selection Guide
TOPSIS	Technique Of ranking Preferences by Similarity to the Ideal Solution
TR	Transportation
RDMF	Remanufacturing Decision-Making Framework
SETAC	Society of Environmental Toxicology and Chemistry
VIKOR	Vlse Kriterijumska Optimizacija Kompromisno Resenje
WEEE	Waste Electrical and Electronic Equipment
US	Usage Stage

List of Symbols

CHAPTER 3

C_{Reman_i}	Remanufacturing cost of component i
$C_{\text{Miscellaneous}_i}$	Cost of handling, disassembly, storage, reassembly of component i
$C_{\text{Shredding/Recycling}_i}$	Shredding/recycling cost of component i
C_{Newpart_i}	Replacement cost of component i
C_{Landfill_i}	Landfill cost of component i
$E_{\text{Material extraction}_i}$	Energy required to extract the raw material to produce component i
$E_{\text{Material process}_i}$	Energy required to process the material to produce component i
$E_{\text{Manufacturing}_i}$	Energy required to manufacture component i
E_{Reman_i}	Energy required to remanufacture component i
$E_{\text{Shredding/Recycling}_i}$	Energy required to shred/recycle component i
RS	Product resale price
CC	Product collection cost
DS_i	Disassembly cost of assembly i
E_{total}	Energy required to produce the product
$X_1 \dots X_m$	Design candidate
$A_1 \dots A_n$	Evaluation criteria
j	Number of the criterion
n	Total number of criteria
α_j	Weight obtained via the entropy method
β_j	Subjective weight assigned by experts from the remanufacturing field
w_j	Weight for the jth criterion
i	Number of the alternative
m	Total number of alternatives
P_{ij}	The normalized performance matrix
r_{ij}	The performance rating of the ith material alternative with respect to the jth evaluation criterion
E_j	The entropy of the normalized values of the jth criterion
α_j	The weight of the entropy of the jth criterion
M	Number of life cycles
μ	Successful remanufacturing rate
i	ith life cycle
$\text{CED}_{\text{Total}}$	CED throughout the entire product life span
CC_{Total}	CC throughout the entire product life span
$\text{CED}_{\text{MEP}_i}$	CED for the material extraction and processing during the ith life cycle
CC_{MEP_i}	CC for the material extraction and processing during the ith life cycle
CED_{MA_i}	CED for the product manufacturing during the ith life cycle
CC_{MA_i}	CC for the product manufacturing during the ith life cycle

CED_{TR_i}	CED for the transportation of product during the ith life cycle
CC_{TR_i}	CC for the transportation of product during the ith life cycle
CED_{US_i}	CED for the use of product during ith life cycle
CC_{US_i}	CC for the use of product during the ith life cycle
CED_{PTB_i}	CED for the take-back of product during the ith life cycle
CC_{PTB_i}	CC for the take-back of product during the ith life cycle
CED_{RE_i}	CED for the product remanufacturing during the ith life cycle
CC_{RE_i}	CC for the product remanufacturing during the ith life cycle

CHAPTER 4

q	Quality level of the product/component
$Pr_{ij}(q1\|q2)$	Probability of the quality of its subcomponent i equals to $q1$, given the quality level $q2$ of assembly j
SA_{ij}	Subassembly ij
$C_{i,j}$	Component $C_{i,j}$
$PF\left(C_{i+1,j}, q_{i+1,j}\right)$	Profit of each component $C_{i+1,j}$ under quality level $q_{i+1,j}$
$PF\left(SA_{ij}, q_{ij}\right)$	The expected profit for processing the subassembly SA_{ij} with quality level $q_{i,j}$
DC_{ij}	Disassembly cost associated with subassembly SA_{ij}
ij	The jth component on ith level
X_{ij}	Indicator of the EOL strategy of component ij
	$X_{ij} = 1$, if the component ij is to be upgraded
	$X_{ij} = 2$, if the component ij is to be restored
	$X_{ij} = 3$, if the component ij is to be discarded with replacement
$\left(EC_{ij}, X_{ij}\right)$	The economic index of component ij, if EOL option X_{ij} is taken
$\left(PC_{ij}, X_{ij}\right)$	The cost of restoring or upgrading the component ij (if $X_{ij} = 1$ or $X_{ij} = 2$) or the cost of disposing the component ij (if $X_{ij} = 3$) (\$)
$\left(NC_{ij}, X_{ij}\right)$	The cost of producing or ordering the new component ij when $X_{ij} = 3$; otherwise $\left(NC_{ij}, X_{ij}\right) = 0$ (\$)
$\left(EV_{ij}, X_{ij}\right)$	The environmental index of component ij, if EOL option X_{ij} is taken
$\left(RMR_{ij}, X_{ij}\right)$	Mass of raw material required for restoring or upgrading the component ij (if $X_{ij} = 1$ or $X_{ij} = 2$) or replacing with a new component ij (if $X_{ij} = 3$) (kg)
RMC_{ij}	Raw material cost (\$/kg)
$\left(EP_{ij}, X_{ij}\right)$	Energy required for restoring or upgrading the component ij (if $X_{ij} = 1$ or $X_{ij} = 2$) or disposal and replacing with a new component ij (if $X_{ij} = 3$) (MJ)
EC	Energy cost (\$/MJ)
$\left(WDP_{ij}, X_{ij}\right)$	Waste generated for restoring or upgrading component ij (if $X_{ij} = 1$ or $X_{ij} = 2$) or disposing the component ij and replacing with a new component (if $X_{ij} = 3$) (kg)
WMC_{ij}	Waste management cost (\$/kg)

$\left(\text{TOX}_{ij}, X_{ij}\right)$ Toxic discharged during restoring or upgrading of component ij
                          (if $X_{ij}=1$ or $X_{ij}=2$) or disposing the component ij and replacing
                          with a new component (if $X_{ij}=3$) (kg)
WMC_{ij} Toxicity management cost (\$/kg)
$\text{PF}\left(C_{ij}, X_{ij}\right)$ The index of component ij, if EOL option X_{ij} is taken

1 Introduction

1.1 SUSTAINABLE PRODUCTION, CIRCULAR ECONOMY AND REMANUFACTURING

Along with the rapid increase in living standards, the consumption of energy and nonrenewable material is rapidly reaching, what many experts believe, unsustainable levels, which poses significant environmental challenges. Given the finite resources of the earth, sustainable production has been widely recognized as the next industrial revolution. It is a concept that requires a holistic approach to close the product life cycle and incorporate different aspects of sustainability throughout a product life cycle (Nasr et al., 2011, Umeda et al., 2012). Recently, the "circular economy" model is also drawing global attention as an approach to decouple economic growth from resource constraints. The underlying principle of the "circular economy" is to make products and materials restorative and regenerative by design and to maintain them at their highest utility and value. As a result, remanufacturing has become one of the key enablers for sustainable production and circular economy, due to its effectiveness in closing the loop on material flows, extending product life cycle and reducing production waste and emission.

Remanufacturing is the process of bringing products back to sound working status, through the process of disassembly, sorting, inspection, cleaning, reconditioning, reassembly and testing, as shown in Figure 1.1 (Lund and Mundial, 1984). Not all the firms engaged in remanufacturing call themselves remanufacturers. Tire remanufacturers call themselves "retreaders", cartridge remanufacturers prefer to use the term "rechargers" and automobile remanufacturers consider themselves "rebuilders". Though different in names, they all share a common nature of bringing a used product to its as-new condition, sometime even surpassing its initial standard.

The benefits of remanufacturing can be summarized as a triple-win situation. The first win goes to the environment, where the used components are diverted from the waste stream to the reusable life cycle and thus lowering the resource and energy consumption, comparing with manufacturing a second new product (Kerr and

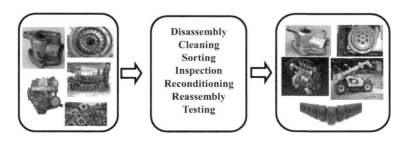

FIGURE 1.1 Remanufacturing processes.

DOI: 10.1201/9781003275596-1

Ryan, 2001). The second win goes to business, as material and energy savings can help companies meet the increasingly stringent environmental legislations, and at the same time preserve the added value from the initial production stage, saving the cost up to 30%–60% as compared to producing a new product (Sprow, 1992). The last win goes to customers because the price of remanufactured products is usually much lower as compared to that of a newly manufactured product (Steinhilper, 1998). The growing awareness of the benefits of remanufacturing has made it an actively perused activity nowadays.

The idea of remanufacturing as an academic research topic began to emerge only in early the 1980s, with Lund and Mundial's original remanufacturing study (1984). Since then, there has been increasing academic interest in remanufacturing arising from its recognized benefits and potential role in changing our society. The decision making on product remanufacturing is a rather complex issue. Business, engineering, market, economic and environmental factors can all affect the success of a remanufacturing endeavor. These important factors have to be investigated before carrying out product remanufacturing or during the product return stage, in order to maintain a profitable remanufacturing business. Moreover, the uncertainties involved in various dimensions, such as quality, quantity and timing of product return, market demand and inventory control, have further complicated the remanufacturing strategy planning issues and make it a challenge to determine the EOL options effectively for a product and its components (Goodall et al., 2014). In this regard, the proposed solutions guide will explore a logical way of determining whether certain products or components are feasible for remanufacturing at the initial product design stage and explore the manner in which a product should be remanufactured during the disposal stage, through weighting and analyzing a comprehensive list of decision-making factors.

Previous research studies have indicated that barriers to the remanufacturing process can be traced to the initial product design stage (Ijomah et al., 2007). Product features and characteristics may have positive or negative impacts on the efficiency of remanufacture, depending upon the decisions made during the design process (Charter and Gray, 2008). These have ignited the concept of design for remanufacturing (DfRem) as a much pursued design activity (Sundin, 2004). The imperative for connecting design and remanufacture is further reinforced by Nasr and Thurston (2006), who stated that the full societal benefits of remanufacturing cannot be achieved unless DfRem is integrated with the product development process. Since design represents one of the earliest product development phases, it is important that the potential for remanufacturing is projected correctly in order that a product can be remanufactured viably and economically. Therefore, the proposed solution will take a proactive measure to improve the potential of a product for remanufacturing, through developing an effective and efficient product design tool to address the remanufacturing issues at the product design stage.

1.2   OVERVIEW OF THE SOLUTIONS GUIDE

The main objective of this solutions guide is to develop a Design for Remanufacturing and Remanufacturability Assessment (DRRA) tool, to be used at the early design stage to provide a systematic and holistic approach toward product EOL decision making, through addressing the comprehensive aspects of remanufacturing issues. The tool also aims to improve the potential of a product for remanufacturing by incorporating remanufacturing considerations into major aspects of the product design. The tool can be used during the product return/service stage to evaluate product remanufacturability based on the return condition and deliver a recovery plan that maximizes the economic profit meanwhile minimizing the environmental impact.

The flow chart in Figure 1.2 shows the focus and scope of the solutions guide. Since the area of product DRRA is fairly wide, there is a need to narrow the scope. The delimitation for this solutions guide is as follows:

a. A complete sustainability problem is built upon three pillars, namely, the economic, environmental and societal pillars. However, due to fact that the social impact evaluation is generally considered to be still in its infancy and can hardly be quantified with a suitable indicator (Klöpffer and Renner, 2008, Jørgensen, 2013, Mattioda et al., 2015), only economic and environmental assessments will be used for the optimization of the remanufacturing decision making.

b. Within this research, case studies have been put forth on automotive products, such as engines and alternators, which have high embedded value, long technology life cycle or high durability, and thus are usually the desired candidates for remanufacturing. However, the case studies also include electrical and electronic goods, such as hedge trimmers and telephones, for which the remanufacturing decision is less clear cut and require an analytical tool for gauging their suitability for remanufacturing. Besides, there are

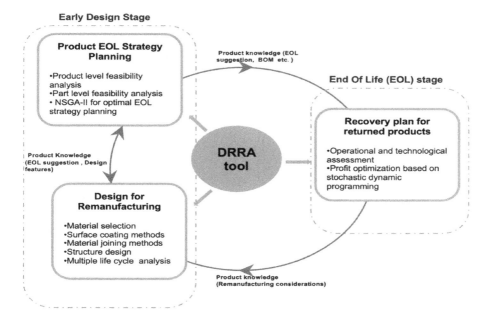

FIGURE 1.2 Flow chart for the solutions guide.

other product categories which might be worth investigating, such as office equipment, medical equipment and aerospace products, which are not covered in the case studies in this book.

c. Remanufacturing know-how, which basically comprises the technology and the information required to remanufacture a product, plays a major role in EOL decision making. The detailed examination and planning of the remanufacturing process and applied technology are not within the scope of this solutions guide. The solutions guide will instead account for factors from a general abstract level.

1.3   CHAPTER OVERVIEW

After this introductory chapter, the next chapter (Chapter 2) summarizes the previous works that are relevant to product DfRem and remanufacturability analysis. The insight gained and limitations identified from these works will constitute the foundation as well as the motivation of this solutions guide. In Chapter 3, the determination of whether a product and its components are feasible for remanufacturing is addressed through the decision-making tool. Meanwhile, the measure of improving the potential of a product/component for remanufacturing will be presented. Besides the early product development stage, the DRRA tool is also developed to be used at the product return/service stage for evaluating product remanufacturability based on the return condition and delivering an optimum recovery plan, and this is discussed in Chapter 4. Chapter 5 discusses the integration of the DRRA tool and introduces the opportunities for Industry 4.0 to support remanufacturing to provide the further perspective of remanufacturing development.

2 Literature Review

2.1 DESIGN FOR REMANUFACTURING

2.1.1 DfRem Activities

Previous design for remanufacturing (DfRem) works have indicated that barriers to the remanufacturing process can be traced to the initial product design stage, and this has ignited the concept of "design for remanufacturing" as a much pursued design activity (Ijomah et al., 2007). The definition of DfRem, as presented by Charter and Gray (2008), is "*a combination of design processes whereby an item is designed to facilitate remanufacture*". DfRem is not only a part of "Design for X" (DfX) methodology, where X represents one of the aims of the methodologies, but it also incorporates a series of DfX strategies, such as design for core collection, design for upgrade, design for disassembly (Charter and Gray, 2008). Sundin (2004) suggested that DfRem stands for a collection of many tasks or considerations which prioritization may vary depending on the processes needed of the products. Soh et al. (2016) proposed a methodology for a systematic, concurrent consideration of design for assembly (DfA) and disassembly guidelines and constraints for product remanufacturing. A set of design feature-based metrics have been proposed by Fang et al. (2014, 2015) for product remanufacturability assessment, namely, disassembly complexity, fastener accessibility, disassemblability and recoverability. Kin et al. (2014) analyzed the conditions of the core components to determine an optimal reconditioning process sequence for design for manufacturing for components. Table 2.1 summarizes the design activities involved in the DfRem methodology.

2.1.2 Desired Product Characteristics for DfRem

Remanufacturing is often practiced by the Original Equipment Manufacturers (OEMs) who remanufacture their own products, contracted remanufacturers who remanufacture the products under contract from the OEMs or customers, or independent remanufacturers (IRs) who buy used products to remanufacture and resell them.

TABLE 2.1
DfRem Activities

• Design for core collection	• Design for restoring
• Design for disassembly/reassembly	• Design for multiple lifecycles
• Design for inspection	• Design for standardization
• Design for cleaning	• Design for handling
• Design for access	• Design for upgrade
• Design for durability	• Design for environment

DOI: 10.1201/9781003275596-2

However, the ability to resolve the difficulties in remanufacturing is most often owned by the OEM, since they control the product design stage and can potentially control remanufacture. Before an OEM considers designing their products for remanufacturing, they should examine whether their products possess the following qualities:

- Product is made up of standard interchangeable parts (Lund and Mundial, 1984).
- The cost of obtaining and reprocessing the core is low compared to the remaining value added (Lund, 1998).
- Technology exists to restore the product (Nasr and Thurston, 2006).
- Product technology is stable over more than one life cycle (Lund, 1998).
- Sufficient customer demand for the remanufactured product (Ayres et al., 1997).
- The core is durable and has high value (Charter and Gray, 2008).
- Potential to be upgraded (Shu and Flowers, 1999).
- There are channels for reverse flow of used products (Ayres et al., 1997).

2.1.3 Guidelines for DfRem

The most commonly used and effective approach to facilitate product DfRem is through providing design guidelines to steer a design toward higher remanufacturability. It is noted that the design guidelines proposed from various literature and research articles have presented a complementary but sometimes overlapping insight. An overview of the design guidelines for successful product remanufacturing is therefore conducted. The collated design guidelines will be presented in a generic and general manner and categorized according to the six steps that constitute the remanufacturing process, namely, core collection, disassembly, inspection and sorting, cleaning, refurbishment, and reassembly and testing. The results can be used to identify the opportunities for enhancing remanufacturing design, set goals and measure progress. Table 2.2 summarizes the literature sources drawn for compiling these guidelines.

2.1.3.1 Design for Reverse Logistic

End-of-Life products usually need to be returned to a specific remanufacturing factory in order for remanufacturing to take place. If this process is not well dealt with, a large cost barrier could occur. For example, to facilitate core collection, the structure should be designed in such a way as to minimize the occurrence of damage during transit. For products which movement requires the use of forklifts, sufficient clearance and support at the base should be provided. In addition, structures that protrude outside a regular geometric volume should be avoided, since they are prone to become damaged during transportation and may also hinder stacking during storage (Shu and Flowers, 1999). Meanwhile, labels, graphical communication and the form of the product should be placed on the exterior or interior surface of the product to communicate the information of the product. For example, radio frequency identification (RFID) is frequently regarded as a form of label to allow a vast array of information to be held (Charter and Gray, 2008).

TABLE 2.2

References Used in Compiling List of Guidelines for DfRem

Reference	Core Return	Disassembly	Sorting and Inspection	Cleaning	Refurbishing	Reassembly and Testing
Amezquita et al. (1995)		✓				✓
Mabee et al. (1999)		✓	✓	✓	✓	✓
McGlothlin and Kroll (1995)	✓	✓	✓	✓	✓	
Shu and Flowers (1999)	✓	✓	✓	✓	✓	
Sundin and Bras (2005)		✓	✓	✓		
Sundin and Lindahl (2008)		✓		✓	✓	✓
Charter and Gray (2008)	✓	✓	✓	✓	✓	✓
Ijomah et al. (2007); Ijomah (2009)		✓	✓	✓	✓	✓
Yüksel (2010)		✓	✓	✓	✓	

2.1.3.2   Design for Disassembly

Disassembly is not a simple reversal of assembly. Many permanent techniques which have been developed to realize and fasten the assembly process, such as plugging, pressing, forming, sonic welding and adhesive, can cause problems for the disassembly process (Mabee et al., 1999). Soh et al. (2014) proposed a conceptual framework to illustrate the integration of methodology, technology and human factors to further enhance the disassembly process. Siew et al. (2020) proposed an integrated framework that considers appropriate human factors and ergonomic conditions in the disassembly process. Liu et al. (2014) proposed a modular design flow to achieve machine tool modularization for remanufacturing. Chang et al. (2017a, b, 2020) proposed a proof-of-concept novel near real-time interactive AR-assisted product disassembly sequence planning system (ARDIS) based on product information, such as interference matrix and 3D models. A detailed review of product disassembly sequence planning has also been conducted by Ong et al. (2021).

Basically, there are four areas that need special attention in design for disassembly:

1. **Joint selection**: the selection of the types of joints would critically affect the efficiency of the disassembly process. Nonpermanent joints are generally preferred over adhesives since they are simple to loosen (Mabee et al., 1999), e.g., bolt joints.
2. **Nondestructive disassembly**: disassembly is desired to be nondestructive (Bras and McIntosh, 1999). After the disassembly, the components are expected to be separated without being damaged or cause damage to other parts of the product. In addition, it is desirable for the fasteners to be reused.
3. **Prevent corrosion/rust**: corrosion and rust are the greatest hindrance reported in the automotive industry survey (Charter and Gray, 2008). Prevention of corrosion and rust using the less or noncorrosive materials or switch to other fastening mechanisms will lead to better isolation of parts from the elements.
4. **Clear instructions for disassembly steps**: the disassembly instructions should be properly displayed on the returned core to facilitate the disassembly process. This is particularly important for third-party remanufacturers, who do not have detailed specifications of the products.

2.1.3.3   Design for Sorting and Inspection

Depending on the various inspection results, parts are sorted into three classes, namely, reusable without reconditioning, reusable after reconditioning and not reusable. To facilitate the sorting and inspection process, parts fulfilling the same function should have identical or distinctly dissimilar features. For example, to differentiate gears that fulfill different functions, gears could be made with color-coding or bar-coding schemes to identify them easily (Mabee et al., 1999). Meanwhile, determining and accessing the point for testing should be made easy and the time required for the inspection of the parts should be minimized. Design features, such as the sacrificial parts for indicating the component's condition over time, should be encouraged. Sensors can also be embedded to record the useful data and communicate the information over time (Ilgin and Gupta, 2011).

2.1.3.4 Design for Cleaning

Cleaning is the most energy- and labor-intensive process in remanufacturing (Shu and Flower, 1999). Therefore, it is important to take the cleaning process into consideration during design; otherwise, the cleaning operation can become too laborious, expensive or even impossible. First, texture and geometrics that facilitate easy cleaning are encouraged, such as a relatively flat surface which has a lower tendency to trap dirt or collect residue from cleaning (Amezquita et al., 1995). Second, structures that require fewer variation of cleaning methods are always preferred. In this way, the cleaning process can be simplified. The material of the product that requires special cleaning methods should be avoided as much as possible, so as to minimize the cleaning cost as well as waste generation (Shu and Flowers, 1999). Third, during the cleaning process, labels and instructions which carry the product information on the component should be prevented from being washed away, since this may cause problems in subsequent refurbishment and reassembly processes (Sundin and Bras, 2005).

2.1.3.5 Design for Reconditioning

During the refurbishment process, parts will be restored geometrically and properties will be restored with surface treatment. To facilitate this process, bulky and slightly overdesigned components are preferred than products with thin and less material, as the former could provide more margin of materials to be worked on with during refurbishment of components (Shu and Flowers, 1999). Surfaces should also be designed in such a way that they have strong wear resistance, since the product may need to go through several use cycles. Moreover, it is appropriate to increase the dimensions to maximize usage cycles since part wear tolerance and material removal must be considered in these areas (Mabee et al., 1999). In addition, a proper incorporation of platform and modularity design can increase the product reusability, through allowing the defunct aspects to be grouped and removed easily while retaining the useful aspects of the product (Charter and Gray, 2008).

2.1.3.6 Design for Reassembly and Testing

Designing products for reassembly and final testing can be improved from the following two aspects. First, during reassembly, the number of the adjustments should be kept low, and adjustments should be easy to make and independent from each other. In addition, the design should be flexible enough to be able to adapt to future technology migration as well as accommodate new configurations of the part.

The lists of design guidelines have provided an understanding of the barriers that may be encountered during remanufacturing processes, as well as directions to enhance the efficiency of product remanufacturing. Appendix I provides more detailed remanufacturing requirements and their related design criteria. Remanufacturing requirements are gathered from the feedback of remanufacturers with respect to improving the efficiency of the remanufacturing process. The design criteria are interpreted and "translated" from the remanufacturing requirements, bringing abstract requirements to concrete design specifications. It aims to provide the product designers with the most comprehensive guidelines to enhance product DfRem. However, the designers may still need to make proper judgment during the design of each individual product.

Though straightforward and comprehensive, the approach of design guidelines for DfRem has been criticized as overly daunting, since it is impossible for designers to consider all these criteria simultaneously and some of the remanufacturing design requirements are intrusive on traditional design (Zwolinski et al., 2006). In addition, there are other issues that the design guidelines do not fully address, such as the subjectivity and customization guidelines (Hatcher et al., 2011).

2.1.4 DESIGN FOR REMANUFACTURING TOOLS

The subsequent development on DfRem focuses on formulating the design tools and methods to address and alleviate the problems associated with the remanufacturing processes during the product design stage.

One of the trends is to develop mathematical models, software tools or statistics references for improving product DfRem, assessing product remanufacturability and prioritization of remanufacturing design criteria, as summarized in Appendix II. Sundin (2004) has developed the "RemPro Matrix", which identifies the relationships between different product properties and specific remanufacturing steps; for instance, the ease of access is closely related to disassembly, cleaning and inspection process. Ijomah et al. (2007) have proposed some fundamental steps required to improve the robustness of the DfRem methodology. However, most of these models and tools still remain within the academic realm and have hardly been utilized in the industry today. Some of the reasons as indicated by Hatcher et al. (2011) are that these design tools are quite complex and lack of applicability to the entire lifecycle of a product. Furthermore, most of these tools are only applicable at the late design stage when most of the decisions have already been made. The reluctance of companies to share their in-house methods, tools and knowledge with the outside world also leads to the barrier between the academic and the industry world.

Another trend of DfRem is to use existing design tools, such as modularization and QFD, which are considered relevant for improving the remanufacturability of products. Appendix III summarizes the design aids that have been used to facilitate DfRem. As most of the designers are familiar with these design tools, this would make the integration of DfRem a much simpler job. However, the problems associated with these tools are that most of them are not developed for DfRem purposes and fail to address all the design aspects that affect the potential of a product for remanufacturing. Therefore, a holistic guidance and assistance on how to carry out DfRem with these tools would need to be further explored.

2.1.5 CHALLENGE AND FUTURE TRENDS OF DFREM

Despite the appealing benefits of carrying out DfRem, there are still barriers and complications that companies may face. First of all, comparing with other DfX issues, such as DfA, DfRem is usually not given the priority, since the main focus of most OEMs is on the manufacturing and usage phases. Whenever there is a conflict between DfRem and other prioritized issues, such as DfA and manufacturing, DfRem usually loses its importance and is viewed as less useful in terms of time and cost due to the lack of awareness among designers. Therefore, a holistic life cycle assessment (LCA) is necessary to quantify the impact of remanufacturing improvement design feature. Second, some OEMs play down on remanufacturing deliberately through product design to stifle the independent remanufacturing activities. This is because none of the OEMs have strong desire to enhance remanufacturability for benefitting the IRs, who are viewed as strong competitors of their own products. Third, DfRem guidelines, as presented in Section 2.1.3, involve a variety of design issues, which will form a new set of challenges that producers may not be prepared to deal with, not to mention there are still confusions around the definition of remanufacturing (Hatcher et al., 2011).

Future research and solution on DfRem can continue to work on developing methods and tools, especially the ones that incorporate life cycle thinking and can be used effectively at the early design stage, as DfRem is most effective in this stage when few design decisions have been made and less technical data is defined (Amezquita et al., 1995; Zwolinski et al., 2006). More case studies, comprising the entire spectrum of the remanufacturable products, are needed for further validation of researchers' findings or design tools. In addition, though the design criteria for remanufacturing have been reviewed comprehensively, the method for integrating them fully into the design process still needs further exploration.

2.2   PRODUCT REMANUFACTURABILITY ASSESSMENT

Among the research works that are related to EOL assessment, two main streams, namely product demanufacturabilty assessment and product remanufacturability analysis, can be identified. Even though both streams deal with the optimization of product EOL disposition, the fundamental differences remain on the final destination of EOL products. Demanufacturing focuses on part-level components' reuse, recycle, remanufacture, landfill or disposal, through dismantling of EOL products (Johnson, 2002; González and Adenso-Díaz, 2005; Jun et al., 2007; Staikos and Rahimifard, 2007; Chan, 2008; Gehin et al., 2008; Zhang et al., 2013). On the other hand, product remanufacturing, which is also the focus of this research, deals with recovering the entire EOL products to as-new condition. The related work on evaluating remanufacturing strategy can usually be observed in two major aspects, product-level remanufacturability assessment and parts-level remanufacturability assessment.

2.2.1   PRODUCT-LEVEL REMANUFACTURABILITY ASSESSMENT

Many solution approaches have been based on the economic benefits to assess the feasibility of product remanufacturing, since remanufacturing without a sound monetary foundation will almost certainly fail (Subramoniam et al., 2009). Decision support tools are necessary to help the decision makers decide whether they should invest in remanufacturing any of their products. Fang et al. (2016) developed a software tool for product remanufacturing assessment, process planning and disassembly route evaluation. King and Barker (2007) have applied the Delphi technique to build a robust research agenda and identified selling "use" instead of "product" as a novel remanufacturing business model. A remanufacturing facility cost model has been developed by Sutherland et al. (2010), which includes product, operation, inventory and transportation-related costs. The output of this work can be used for facility planning for remanufacturing operations. Chen and Chang (2012) have built an economic model to analyze the pricing and production lot-sizing in a closed-loop supply chain and used this model to investigate the possibility to combine remanufacturing with manufacturing operations. A cost model has been developed by Xu and Feng (2014) to evaluate the benefits of remanufacturing techniques quantitatively and assist decision making on EOL strategies. However, since the products could be returned multiple times, it is possible that the material, labor and overhead costs could only be recaptured and a profit made after several sales, which makes the determination of the profit of remanufacturing even more complicated. The transfer of pricing to allocate a portion of the initial production cost to the remanufacturing division has been discussed by Toktay and Wei (2011), aiming to achieve the optimal financial results of the firm. However, mere focus on economic feasibility of product remanufacturing could lead to inadequate support for remanufacturing decisions and result in sub-optimization of the entire supply chain, since there are other factors, such as ecological factors and business factors that are also influential on remanufacturing decisions. Therefore, some studies have focused on developing decision-making frameworks, which are comprised of a comprehensive set of strategic factors, such as customer demands and environmental consideration. For example, a software tool has been established by Kobayashi (2005) to assign the appropriate life cycle options to the product and its components, taking into account the business, production and environment perspectives. Remery et al. (2012) have utilized the Fuzzy TOPSIS method to assist the multicriteria EOL decision making. Subramoniam et al. (2010) have used survey ranking to prioritize 12 deciding factors for remanufacturing, and built a comprehensive Remanufacturing Decision-Making Framework (RDMF). Other principal operation control issues that hinder remanufacturing have been identified by Ijomah (2009), which include the uncertainty of demand volume variability, core quality, the difficulty of knowledge acquisition and process as well as the flexibility issue. In addition, some commercial tools have been made available for remanufacturing assessment and supply chain management (Levelseven, 2015; Activate, 2015; NCMS, 2015; Ipoint, 2015).

With the awareness that design determines two thirds of the product remanufacturing efficiency, some researchers have started to assess product remanufacturability specifically from the design perspective. Lund (1998) has proposed seven major criteria for product remanufacturability assessment at the design stage, based on the study of 75 routinely remanufactured product types. Amezquita et al. (1995) have consolidated design metrics to measure effectively and efficiently the remanufacturability of product design, which includes design characteristics that facilitate remanufacturing, the principal driving factors for remanufacturing as well as the existing remanufacturing guidelines and practices. Adapted from DfA metrics, Bras and Hammond (1996) have developed the metrics for assessing the remanufacturability of a designed product in a qualitative way. Besides, an integrative approach to assess the technical, economic and environmental feasibility of the returned products and facilitate the decision making on whether a product should be remanufactured, has been proposed by Du et al. (2012) and validated in a machine tool remanufacturing study.

2.2.2 PARTS-LEVEL REMANUFACTURABILITY ASSESSMENT

While the literature and theories focusing on strategic decision making for product remanufacturing are gaining popularity, the available framework on determining the EOL strategies at the component level during the product service stage is relatively limited. Even for the routinely remanufactured products, not all their components are suitable for remanufacturing. Some components are disposed of due to severe wear and corrosion, while others are recycled to recover the raw material (Smith and Keoleian, 2004). Thus far, decision making on components' EOL strategy planning relies mostly on *ad hoc* engineering judgment, which may be subjective and imprecise. Therefore, there is a need for systematic and comprehensive decision support tools to evaluate remanufacturability at the component level so as to help the decision makers make better EOL choices.

Disassembly, which allows the separation of the reusable and non-reusable components for further processing, is closely related with EOL strategy determination and regarded as a new frontier to product EOL management. Many researchers have proposed different methodologies to measure the disassemblability of a product and generate an optimum disassembly sequence. For example, Gungor and Gupta (1997) have proposed a disassembly sequence generation heuristic which could generate the optimum disassembly sequence for a product. Pbioore et al. (1998) have used Petri Nets to study disassembly planning. Differences among these methods are the techniques applied to solve the problems. Moreover, the factors that affect the EOL strategy include not only the disassembly sequence, the disassembly time, the disassembly cost, but also the benefits from reusing and recycling the components (Veerakamolmal and Gupta, 1999). There is, therefore, a growing amount of work on proposing methods for generating "recovery plans" and balancing the value of the reclaimed parts with the disassembly cost. Isaacs et al. (1997) have proposed a methodology to measure the disassembly and recycling potential for automobile design. González and Adenso-Díaz (2005) have introduced a model which could determine the optimal EOL strategy for each component and the subsequent disassembly strategy that leads to the highest profits.

With the increasingly restrictive environmental regulations, it is not only critical to maintain economic profitability, but also crucial to minimize the environmental impact through product life cycles, which has led to the multiobjective decision-making problems (Hula et al., 2003). Traditionally, these two objectives were either combined linearly to form a scalar objective or only one objective is optimized and the other one is turned into a constraint; for example for the former, Ghazalli and Murata (2011) have converted environmental impact into environmental cost and integrated it with economic cost to determine the component EOL strategies; for the latter, the EOL decision model provided by Lee et al. (2010) optimizes the economic profit and used environment regulation as a constraint to revise the EOL options.

2.2.3 CHALLENGE AND FUTURE TRENDS OF REMANUFACTURABILITY ASSESSMENT

The decision making on a product and its components remanufacturing is a rather complex issue. Business, engineering, market, economic and environmental factors can all affect the success of a remanufacturing endeavor. These important factors have to be investigated before carrying out product remanufacturing and are crucial for maintaining a profitable remanufacturing business. Moreover, the uncertainties involved in various dimensions, such as time and quantities of product return, market demand and inventory control, have further complicated the remanufacturing strategy planning issues and make it a challenge to determine the EOL options effectively for a product and its components (Goodall et al., 2014). In this regard, there is an urgent need of a decision support tool that can provide a systematic and holistic approach toward EOL decision making, through addressing comprehensive aspects of remanufacturing issues as well as weigh in on the uncertainty involved in product remanufacturing process. Furthermore, future research can focus on developing databases or knowledge-based systems, drawing experience from the existing remanufacturing knowledge and practices, to facilitate the new assessment tools.

2.3   RELATED METHODOLOGIES AND APPROACHES

2.3.1   DESIGN FOR ENVIRONMENT

DfE is to design a product such that the environmental impact throughout the life cycle is minimized (Ilgin and Gupta, 2010). Some researchers adopted QFD methodology to consider environmental criteria and customer requirements simultaneously for product design (Cristofari et al., 1996; Zhang, 1999; Mehta and Wang, 2001). Other researchers use LCA to assess the environmental impact (Veerakamolmal and Gupta, 1999; Grote et al., 2007). There are studies that focus on developing tools to evaluate product design with respect to environmental criteria, such as Green Design Advisor, which considers metrics related to product design information and combines these metrics using the multiattribute value theory to obtain an overall score (Feldmann et al., 1999). Lye et al. (2002) proposed EcoDe, which is a computer-based design evaluation tool, to assess the environmental impact of the components of a product. Analytic hierarchy process (AHP) and the multicriteria technique are used in this tool to calculate the environmental index.

DfRem is generally viewed to be under the umbrella of DfE. Compared with DfE, DfRem is a relatively new and unexplored research area. Thus, the literature on DfE provides a valuable insight on the approaches that are likely to be applicable in DfRem. For example, DfE literature emphasizes the importance of early integration of environmental requirements, the positive impact from management commitment and the indispensability of tools to address the environmental requirements, which inspire the ways toward successful DfRem implementation (Quella and Schmidt, 2003). However, DfE and DfRem are not interchangeable, and sometimes, they may even be in conflict with each other. For example, DfRem may require components to be overdesigned such that in subsequent remanufacturing operations, e.g., machining and grinding, can be performed easily; on the other hand, DfE may require components to be designed with minimum use of materials so as not to waste resources. The difference between them emphasizes the importance of exploring DfRem as a standalone entity.

2.3.2 LIFE CYCLE ASSESSMENT

To evaluate and assess the environmental impact attributed to the life cycle of a product and identify improvement potential, a large number of assessment methodologies and corresponding indicators have been developed (Hertwich et al., 1997; Robèrt et al., 2002; Umeda et al., 2012; Kobayashi, 2006, Durairaj et al., 2001, 2002, 2003). Among these proposed tools, LCA methodologies are the most widely used. Its underlying philosophy is to provide a comprehensive view of the environmental aspects of a product or process throughout the life cycle and an accurate analysis of the environmental trade-offs in product and process selection (Corporation and Curran, 2006). Figure 2.1 depicts a framework for LCA by the Society of Environmental Toxicology and Chemistry (SETAC) (Rebitzer et al., 2004). Problems, such as shifting from one stage of the life cycle to another, from one type of problem to another, and from one location to another, can be avoided through this integrative approach, since all the life cycle stages are included in the evaluation (Zhang, 1999; Mehta and Wang, 2001; Grote et al., 2007; Sakao, 2007).

However, one of the limitations of carrying out an LCA is that a significant amount of data is required. Even though some software programs with inventory data are available, gathering data for specific product processes still remains a challenge, as most of these data are not available to the public or not provided in a standard format (Huijbregts et al., 2006). Generally, only little information on product or process is available during early product development stage, which has limited the applicability of LCA in the early design stage. Furthermore, as remanufacturing can close the loop of material flow effectively and has great potential to increase the number of product life cycle from one to multiple, the resulting complex life cycles have to be modeled and assessed by decision makers for a better understanding of the relative

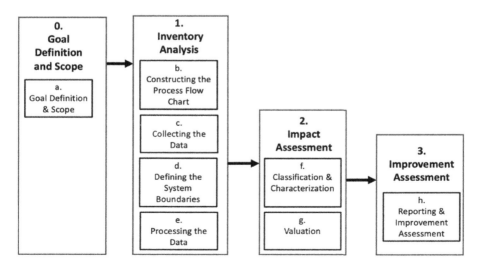

FIGURE 2.1   LCA framework (Rebitzer et al., 2004).

benefit of the alternative design strategies, and thus allowing them to make decisions on improving product design and EOL recovery system.

2.3.3 MULTIOBJECTIVE DECISION-MAKING ANALYSIS

Making an optimum decision for diverse applications from a finite set of feasible alternatives and predetermined number of criteria is often regarded as a multicriteria decision-making (MCDM) problem. MCDM methods, such as Technique Of ranking Preferences by Similarity to the Ideal Solution (TOPSIS) (Jee and Kang, 2000; Shanian and Savadogo, 2006; Rao, 2008; Zeydan and Çolpan, 2009; Govindan et al., 2013), elimination and choice expressing the reality (ELECTRE) (Milani and Shanian, 2006; Shanian and Savadogo, 2006; Shanian et al., 2008; Laforest et al., 2013), AHP (Cao et al., 2006; Dweiri and Al-Oqla, 2006; Rao, 2008; Jiang et al., 2011), preference ranking organization method for enrichment evaluation (PROMETHEE) (Chatterjee and Chakraborty, 2012; Çalışkan et al., 2013; Peng and Xiao, 2013), Vlse Kriterijumska Optimizacija Kompromisno Resenje (VIKOR) (Opricovic and Tzeng, 2004; Opricovic, 2011; Shemshadi et al., 2011) and simple additive weighting method (Quigley et al., 2002; Dehghan-Manshadi et al., 2007; Fayazbakhsh et al., 2009) are the most notable. Though each technique has its own characteristics, they share a similar systematic evaluation procedure which involves the following three steps (Yurdakul and Ic, 2009):

1. Determine the relevant criteria and feasible alternatives,
2. Attach numerical measures to the relevant importance of the criteria considered and the impact of the alternatives of those criteria,
3. Determine a ranking score of each alternative by processing the numerical values.

Among these MCDM methods, TOPSIS is chosen in this research to evaluate and compare the feasibility of design candidates. The TOPSIS method is proposed by Hwang and Yoon (1981) based on the idea that the best alternative should have the shortest distance from an ideal solution and farthest from the nonideal solution. It is a widely used MCDM tool due to its simplicity and effectiveness. The output of this method can be a preferential ranking of the alternatives with numerical values (Shanian and Savadogo, 2006). The application of TOPSIS in the product design stage has been observed from many published works. For example, Shanian and Savadogo (2006) have applied the TOPSIS analysis to select optimum materials for metallic bipolar plates for a polymer electrolyte fuel cell. Boran et al. (2009) have selected the suppliers successfully, who are able to provide the buyer with the right quality products and/or services at the right price, at the right time and in the right quantities through the use of the TOPSIS methodology.

The traditional TOPSIS method defines the problem in the form of a decision matrix filled with crisp data, assuming that the performance value is defined precisely. However, in some real-world decision-making situations, due to time pressure and limited information or knowledge about the problem domain, decision makers may prefer to express their evaluation qualitatively with verbal descriptions or linguistic variables, rather than exact numbers. Therefore, some researchers have proposed to combine TOPSIS with the fuzzy set theory for expert evaluation and adopted this Fuzzy TOPSIS method in the area of decision making on green supply chain (Wang and Chan, 2013a), supplier or outsourcing manufacturing partner selection (Chen and Hung, 2010; Govindan et al., 2013), plant location selection (Ertuğrul and Karakaşoğlu, 2008), risk assessment (Samvedi et al., 2013), material selection and design (Rathod and Kanzaria, 2011; Mirhedayatian et al., 2013), etc.

2.4   SUMMARY OF THE CHAPTER

This chapter has provided an overview of the existing DfRem approaches. The problems of existing DfRem tools and guidelines, such as overly daunting, lack of life cycle thinking and overly complex, have been identified to address the need of a DfRem tool that can be used effectively at the early design stage and integrated easily with the original design process. On the other hand, research works that are related to product-level and part-level remanufacturability assessments have been reviewed in this chapter. Though diverse and informative, there are still some limitations of the existing approaches, such as a lack of a holistic and systematical tool that can weigh in on comprehensive remanufacturing considerations while achieving both economic and environmental optimizations. Besides, EOL decision making at the early development stage and product return stage presents different problems sets, such as the availability of product information, which most of the research work did not mention much and distinguish between them. Hence, there is a need of tools to be developed specifically to facilitate the decision making at both of these stages. Overall, the insight gained and limitation identified from this literature will constitute to the foundation as well as the motivation of this solutions guide.

3 Early Design Stage

3.1 EOL STRATEGY PLANNING

3.1.1 INTRODUCTION

The decision to include remanufacturing as a part of product design should be made as early as possible. However, this decision-making process is a rather complex issue as it involves high level of uncertainties of time and quantities of product return, market demand, inventory control, etc. Even though some relevant work has been identified, there are still limitations from the following viewpoints. First, as the research work on decision support for remanufacturing strategy planning is still in its infancy, there is a lack of a holistic approach that can address the various aspects of remanufacturing considerations and ensure the completeness of financial and environmental decisions. Second, given the multi-criteria nature of EOL strategy planning, the solution is usually a range of possible choices rather than an exact one. A decision method has yet to be described that can determine a set of optimum remanufacturing strategies efficiently and quantitatively and provide more flexibility for remanufacturing strategy planning. Third, due to the uncertainty that is involved in the EOL strategy planning process, the decision model that can analyze the impact of situational variables effectively on EOL decision making has not been found in any literature.

In this regard, a methodology and decision support tool dealing with remanufacturing strategy planning are presented in this chapter. The methodology and decision support tool provides a systematic and holistic approach toward EOL decision making, through addressing comprehensive aspects of remanufacturing issues. A Genetic Algorithm, namely, NSGA-II, has been adopted in the decision support tool to determine a Pareto set of optimal EOL solutions; the algorithm will facilitate the effort of decision makers to maximize the environmental benefit of remanufacturing for a given economic profit. In addition, the rapid calculation of Pareto solutions through the methodology permits extensive sensitivity analysis so as to understand thoroughly the impact of situational variables on EOL decision making, i.e., Pareto frontier, and thus leading to improved strategy planning and better product design. Hence, the output of the methodology is essential for making the following decisions:

- Evaluating the feasibility of a product and its subassemblies/components for remanufacturing and redesigning the process and/or product if necessary;
- Determining a Pareto set of optimum solutions corresponding to maximum environmental performance for a given economic cost as well as suggesting EOL strategies for subassemblies/components;
- Determining the economic viability of the depth or extent of disassembly during remanufacturing;
- Investigating a large number of scenarios as necessary in EOL decision making (Pareto frontier), such as change of remanufacturing cost, product

DOI: 10.1201/9781003275596-3

design and landfill cost, to suggest appropriate EOL strategies to accommodate the change of situational variables.

3.1.2 FRAMEWORK FOR PRODUCT EOL STRATEGY PLANNING

To assist EOL decision making, especially for companies which are carrying out remanufacturing or planning to engage in remanufacturing business, a four-step decision support tool is presented and shown in Figure 3.1. The first step consists of a sequential examination of product-level remanufacturing characteristics, aiming to distinguish quickly a product that is feasible for remanufacturing from a product that is not a viable candidate for remanufacturing. Next, a detailed subassembly-/component-level characteristic examination will be carried out to identify the viable subassemblies/components for remanufacturing. In the third step, a multi-criteria decision-making (MCDM) analysis is conducted to generate a Pareto set of EOL solutions, from which decision makers can choose to accommodate various remanufacturing requirements. The last step will incorporate sensitivity analysis to examine the impact of situational variables or product redesign on EOL strategy planning, thus leading to improved strategy planning and better product design. The details of the methodology are presented next (Yang et al., 2016).

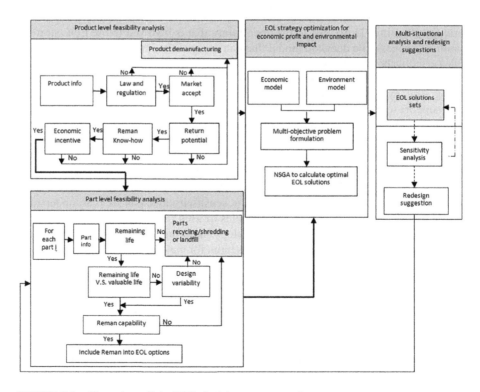

FIGURE 3.1   Flow chart of the EOL decision support tool.

3.1.2.1 Step I: Product-Level Feasibility Analysis

The product-level feasibility analysis is developed to distinguish a product that should be designed for remanufacturing quickly from a product that should be targeted for the demanufacturing strategy, through a sequential examination of characteristics that are related to product remanufacturing performance, as elaborated next.

- **Compliance with laws and regulations**: The impact from environmental legislations varies for product remanufacturing. On the positive side, these regulations address the necessity of sustainable development and thus promote remanufacturing development. However, some of them may hamper remanufacturing to various degrees. For example, the European Commission's Restriction of Hazardous Substances (RoHS) Directive may hinder the markets for remanufactured products that contain substances now termed "hazardous" (Charter and Gray, 2008). As mentioned by Gerrard and Kandlikar (2007), the hazardous substance concentration from the European Commission's End-of-Life Vehicles (ELV) Directive on recycling and energy recovery may discourage higher forms of waste management hierarchy, such as remanufacturing. Therefore, the impact from legislations on remanufacturing needs to be understood in order to design successful products for remanufacturing.
- **Market demand/acceptance**: This criterion evaluates the extent to which a design concept is expected to occupy a competitive position in the market after being remanufactured. For product types that have a long technological cycle, like the diesel engine, heavy-duty vehicles, etc., remanufacturing is usually feasible since most of those products can still remain competitive in the aftermarket. For product types that have rapid technology obsolescence, remanufacturing might not be a feasible EOL option, unless the design features that are likely to suffer early obsolescence can be decoupled easily from the more stable product platforms, or secondary and tertiary markets, which usually do not require the product to possess the latest technology, can be found.
- **Return potential**: It is important to predict the likelihood that a product will be returned successfully to the remanufacturing site at the end of its useful life. Some remanufacturing companies purchase retired products from collection points in the existing distribution and disposal networks, or directly from customers; some OEMs establish leasing arrangements to retain the product ownership and have products returned at a specified time; others incorporate "core charge" to their products and reimburse the customers only when the products are returned properly, etc. Failing to identify the product return channels may cause infeasibility in carrying out product remanufacturing.
- **Remanufacturing know-how**: Remanufacturing know-how comprises the technology and the information required to remanufacture a product. Decision makers need to examine whether the technology and technical skills needed to restore a product will be available when the product reaches

its EOL stage, such as the technology to repair wear in stressed areas of cast iron engine blocks, the skill to disassemble a sophisticated joint. Meanwhile, the availability of the technical data package to restore a product needs to be examined, which includes material specifications, dimensional tolerance and historical information that stores the product refurbishment history.

- **Economic incentives**: Economic incentives come from both product value and product recoverable value. Decision makers need to determine whether a product has undergone sufficient value-added operations (e.g., manual or automated operations) to make remanufacturing worthwhile. On the other hand, they also need to estimate whether a product can be economically recoverable at its EOL stage.

Hence, if a design concept performs satisfactorily for all the criteria, it is considered as a feasible candidate for remanufacturing and will enter into the second stage for part level evaluation. Failing to meet one or more of the criteria will indicate that remanufacturing is not practical. In this case, demanufacturing will be the fallback position and the remanufacturing strategy planning process will stop.

3.1.2.2   Step II: Part Level Feasibility Analysis

Even though a product is suitable for remanufacturing, not all its components can be remanufactured. Therefore, the purpose of this stage is to determine the feasible EOL options for each subassembly/component and rule out the ones that are not viable for remanufacturing.

- **Remaining useful lifetime**: Useful lifetime is defined as the time from product/part purchase until the product/part no longer meets its initial requirement, due to "failure" or "physical degradation" (Rose, 2000). The remaining useful lifetime of a component after being remanufactured should correspond to more than at least one usage period, such that the quality of the remanufactured product can be ensured.
- **Remaining useful lifetime versus valuable lifetime**: Valuable lifetime refers to the time that a product/part is expected to occupy a competitive position in the marketplace, until it becomes obsolete or less desirable due to external factors, such as market pressure, scientific advances and company focus. If the valuable lifetime is longer than its useful lifetime, the part will usually have high remanufacturing potential. However, if the valuable lifetime is shorter than the useful lifetime, the part will be evaluated for the next criterion for design viability examination.
- **Design viability**: This evaluation criterion is meant to encourage designers to rethink the design concept in order to prolong the valuable lifetime of a part, such as incorporating modularity design to facilitate upgrade, replace or other forms of enhancement. If the review suggests negative on design viability, recycling or landfill of the part will be suggested.
- **Remanufacturing capability**: Remanufacturing capability will be assessed from the availability of facility, equipment and trained personnel to perform

the remanufacturing procedures. When internal resource is insufficient to meet remanufacturing requirements, feasibility to outsource the parts for remanufacturing might also be explored.

Therefore, if subassemblies/components can pass the feasibility assessment successfully, remanufacturing will be included as one of their feasible EOL solutions, otherwise only shredding/recycling and landfill will be considered. Once all the feasible EOL strategies have been identified for each subassembly and component, the evaluation process will advance to step III.

3.1.2.3 Step III: EOL Strategy Optimization for Economic Profit and Environmental Impact

To optimize the EOL strategy planning, the value of economic profit and environmental benefit needs to be estimated first using Equations 3.1–3.5. It is noted that energy has been chosen as the indicator for environmental performance, as it has strong correlation with the various environmental metrics, such as global warming potential, air pollutants emissions (Hula et al., 2003). The overall economic and environmental metrics calculated using Equations 3.6 and 3.7 represent the objective functions for the optimization problems. Both objectives are functions of X_{ij}, which represents the EOL strategy j assigned for component i.

The fundamental constraints in this optimization problem include (a) there is only one EOL option for each subassembly or component within the parent assembly, namely, (1) remanufactured, (2) shredded/recycled, (3) landfilled, (4) disassembled or (5) remained; (b) if assembly k is remanufactured, shredded/recycled or landfilled as a complete entity, all of its subcomponents i should remain within this assembly; and (c) if assembly k is to be disassembled, all of its subcomponents i will be separated for remanufacturing, shredding/recycling or landfill. These constraints are expressed in Equations 3.8–3.10.

$$C_{\text{Overall\_Reman}_i} = C_{\text{Reman}_i} + C_{\text{Miscellaneous}_i} \tag{3.1}$$

$$C_{\text{Overall\_Shred/Rc}_i} = C_{\text{Shredding/Recycling}_i} + C_{\text{Newpart}_i} + C_{\text{Miscellaneous}_i} \tag{3.2}$$

$$C_{\text{Overall\_Lf}_i} = C_{\text{Landfill}_i} + C_{\text{newpart}_i} + C_{\text{Miscellaneous}_i} \tag{3.3}$$

$$E_{\text{Overall\_Reman}_i} = E_{\text{Material extraction}_i} + E_{\text{Material process}_i} + E_{\text{Manufacturing}_i} + E_{\text{Reman}_i} \tag{3.4}$$

$$E_{\text{Overall\_Shred/Rc}_i} = E_{\text{Material extraction}_i} + E_{\text{Material process}_i} + E_{\text{Shredding/Recycling}_i} \tag{3.5}$$

$$C_{\text{Total}} = C_{\text{Resale}} + C_{\text{Collection}} + \sum_{i \in \text{reman}} C_{\text{Overall\_Reman}_i} * X_{i1} + \sum_{i \in \text{Shred/Rc}} C_i * X_{i2}$$

$$+ \sum_{i \in \text{Lf}_i} C_{\text{Overall\_Lf}_i} * X_{i3} + \sum_{i \in \text{DS}} C_{\text{Disassemble}_k} * X_{i4} \tag{3.6}$$

$$E_{\text{Total}} = \left(\sum_{i \in \text{reman}} E_{\text{Overall\_Reman}_i} * X_{i1} + \sum_{i \in \text{Shred/Rc}} E_{\text{Overall\_Shred/Rc}_i} * X_{i2} \right) \Big/ E_{\text{Production}} \qquad (3.7)$$

Subject to:

$$\sum_i (X_{i1} + X_{i2} + X_{i3} + X_{i4} + X_{i5}) = 1 \quad X_{i1}, X_{i2}, X_{i3}, X_{i4}, X_{i5} = 1 \text{ or } 0 \qquad (3.8)$$

For subcomponent i that belongs to assembly k:

$$\text{If } X_{k1} + X_{k2} + X_{k3} + X_{k5} = 1; \quad \text{then } X_{i5} = 1 \qquad (3.9)$$

$$\text{If } X_{k4} = 1; \quad \text{then } X_{i1} + X_{i2} + X_{i3} = 1 \qquad (3.10)$$

where:

$C_{\text{Overall\_Reman}_i}$: economic profit obtained from remanufacturing component i;

C_{Reman_i}: remanufacturing cost of component i, which can be estimated from the multiple of labor cost and the total time required for remanufacturing component i, which includes sorting, cleaning, reconditioning and testing process;

$C_{\text{Miscellaneous}_i}$: cost of handling, disassembly, storage and reassembly of component i;

$C_{\text{Overall\_Shred/Rc}_i}$: economic profit obtained from shredding/recycling component i and replacing with new parts;

$C_{\text{Shredding/Recycling}_i}$: shredding/recycling cost of component i, which can be estimated from the scrap value of the materials times the weight of the material;

C_{Newpart_i}: replacement cost of component i;

$C_{\text{Overall\_Lf}_i}$: economic profit obtained from landfilling of component i and replacing with new parts;

C_{Landfill_i}: landfill cost of component i, which can be estimated from the landfill cost times the weight of the material;

$E_{\text{Overall\_Reman}_i}$: energy recovered from remanufacturing component i;

$E_{\text{Overall\_Shred/Rc}_i}$: energy recovered from shredding/recycling component i;

$E_{\text{Material extraction}_i}$: energy required to extract the raw material to produce component i;

$E_{\text{Material process}_i}$: energy required to process the material to produce component i;

$E_{\text{Manufacturing}_i}$: energy required to manufacture component i;

E_{Reman_i}: energy required to remanufacture component i;

$E_{\text{Shredding/Recycling}_i}$: energy required to shred/recycle component i;

C_{Total}: overall economic profit for EOL disposition strategy;

E_{Total}: percentage of energy recovered for EOL disposition strategy;

C_{resale}: product resale price;

$C_{\text{Collection}}$: product collection cost;

$C_{\text{Disassemble}_k}$: disassembly cost of assembly k;

$E_{\text{Production}}$: energy required to produce the product, including material extraction energy, material processing energy and manufacturing energy.

Solving these multi-objective functions requires a discrete optimization algorithm due to their combinatorial nature. A simple enumeration is computationally too expensive even for simple products with relatively small number of components. Thus, the NSGA-II is chosen to approximate the optimum trade-off solutions between the economic profit and the recovered energy rapidly. In NSGA-II, the chromosomes are codified in the forms of a string consisting of $(N+M)$ genes, where N and M represent the numbers of the subassemblies and components, respectively. The value of each gene can be an integer value from 1 to 5, which represent five different EOL options, namely, remanufactured, shredded/recycled, landfilled, disassembled and remains within the parent assembly. The implementation details for NSGA-II can be found in Deb et al. (2002).

3.1.2.4   Stage IV: Multi-Situational Analysis and Redesign Suggestions

The result obtained from NSGA-II is a Pareto set of EOL solutions based on trade-offs between economic profit and environmental impact, subject to the three constraints presented. As the Pareto sets of EOL solutions can be generated within a reasonable time, the algorithm thus permits extensive sensitivity analysis to understand thoroughly the impact of situational variables, such as landfill cost, labor cost, collection cost, remanufacturing capability, on EOL decision making (Pareto frontier). Meanwhile, redesign ideas or solutions can be tested and verified efficiently to promote better EOL strategy planning and product development.

3.1.3   CASE STUDIES

Two types of EOL desktop phones have been chosen as case studies, namely, a normal consumer desktop phone and a business IP desktop phone. Desktop phone remanufacturing can be dated backed to more than 50 years ago, aiming to give a second life to the used and defective desktop phone equipment. Recently, manufacturers and producers of desktop phones are also given greater responsibility under the European Commission Waste Electrical and Electronic Equipment (WEEE) Directive for the collection, recovery or recycling of the e-waste that their goods become, if they produce electrical and electronic equipment or import them into Europe. Due to the limited information available, it is assumed that both desktop phones and their components have fulfilled the remanufacturing feasibility requirements, implying that steps I and II of the decision support tool will not be included in this case study. This assumption will be further examined in the next section through sensitivity analysis to address the uncertainties that may be involved. The economic information as well as the bill of material of the desktop phones was taken from Johnson (2002), which includes product resale price, part remanufacturing cost, disassembly/reassembly cost, shredding/recycling cost, new part cost and landfill cost for each subassembly/component. For environmental impact analyses, the following data and assumptions are adopted:

1. Energy consumption for material extraction and processing (MEP) is approximated by the embodied energy of materials. The database for material embodied energy is obtained from Curlee et al. (1994).

2. As the energy intensity of conventional manufacturing processes for metals, plastics and many composites fall roughly within the range of 1–30 MJ/kg, the manufacturing energy intensity for each component is assumed to be 15 MJ/kg (Duque Ciceri et al., 2010). For electronic components, which have relatively high energy intensity, the method to estimate their manufacturing energy will be adopted from Ashby (2012) and Kemna et al. (2005).

3. Energy consumption of a remanufacturing process can be expressed through the ratio of remanufacturing energy consumption to the original manufacturing energy consumption. Usually, the ratio ranges between 2% and 25%, subject to different conditions (Sutherland et al., 2008). In this case, the value 25% will be taken to make a conservative estimate of energy saving through remanufacturing.

4. The energy required to reprocess the materials at their EOL stage is called "secondary material production energy". The estimated values of the secondary production energy for different materials are provided by Curlee et al. (1994) and Sullivan and Hu (1995).

Figures 3.2 and 3.3 show the graphical representations of both types of phones. Using the cost information, bill of material as well as the energy consumption data, the EOL economic and environmental profits for each subassembly and component have been calculated and shown in Tables 3.1 and 3.2. Using the energy recovered through remanufacturing for component C2 of the consumer desktop telephone as an example, the value is calculated as the sum of the MEP energy (7652.46 kJ), manufacturing energy (1455.00 kJ) and remanufacturing energy (−363.75 kJ). These values will be

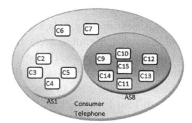

AS1: Handset; C2: Upper Phone; C3: lower Phone; C4: Electrical Subassembly; C5: Speaker; C6: Electric cord 1; C7: Electric Cord 2; AS 8: Phone base; C9: Upper Shell; C10: Circuit Board; C11: Al casting1; C12: Al casting 2; C13: Silence switch; C14: Hang up switch; C15: Lower Shell

FIGURE 3.2   Graphical representation of consumer desktop telephone.

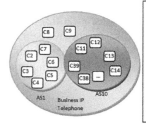

AS1: Handset; C2: Upper Handset; C3: Lower Handset; C4: Handset Circuit Board; C5: Connector; C6: Receiver; C7: Microphone Assembly; C8: Electrical Cord 1; C9: Electrical Cord 2; AS10: Telephone base; C11: Upper Plastic Casing; C12: Speaker Foam; C13:4 Screws; C14: Speaker; C15: Plastic Plate; C16: PolyFlex Circuit Board; C17: LCD Panel#1-Subassembly; C18: Main Circuit Board; C19: Elastomer Key Pad#1; C20: Elastomer Key Pad#2; C21: Elastomer Key Pad#3; C22: Shift Button; C23: 12 Memory keys; C24: 3 Display Buttons; C25: Feature Buttons; C26: Release button; C27 Hold Button; C28: 12 Dial Keys; C29: 12 Line Keys; C30: Volume Control; C31:Display Cover-LCD#1; C32: Display Cover-LCD#2; C33: LCD Panel #2; C34: Hang Up Level; C35:Metal Clip; C36: Lower Plastic Casing; C37: 2 Rubber Pads; C38: Plastic Stand; C39: 2 Rubber Pads

FIGURE 3.3   Graphical representation of business IP telephone.

TABLE 3.1

Economic and Environmental Data for Consumer Telephone

Name	Eco_Reman ($)	Eco_Shrd/Rc ($)	Eco_Lf ($)	Eco_Dis ($)	Evn_Reman (kJ)	Evn_Shrd/Rc (kJ)
AS1	$-\infty$	$-\infty$	−3.12	−0.167	0.00	0.00
C2	−0.85	−1.13	−1.13	$-\infty$	8743.71	2666.27
C3	−0.76	−1.23	−1.22	$-\infty$	8743.71	2666.27
C4	−1.44	−0.43	−0.52	$-\infty$	4214.28	2384.60
C5	−0.12	−0.18	−0.20	$-\infty$	7532.68	4557.02
C6	−1.04	−0.22	−0.24	$-\infty$	5066.73	3032.55
C7	−1.09	−0.26	−0.28	$-\infty$	5066.73	3032.55
AS8	$-\infty$	$-\infty$	−4.68	−1.25	0.00	0.00
C9	−0.84	−0.90	−0.91	$-\infty$	26,377.38	7330.56
C10	−2.47	−1.11	−1.21	$-\infty$	81,130.16	19,495.48
C11	−0.17	−0.32	−0.37	$-\infty$	9099.62	5230.17
C12	−0.17	−0.32	−0.37	$-\infty$	9099.62	5230.17
C13	−0.49	$-\infty$	−0.22	$-\infty$	4388.73	0.00
C14	−0.33	−0.41	−0.41	$-\infty$	3510.98	1004.82
C15	−1.12	$-\infty$	−1.14	$-\infty$	26,332.38	0.00

used as input for the optimization model to determine the optimum set of EOL strategies. The minus infinite signs in Tables 3.1 and 3.2 indicate the infeasible EOL options.

The consumer telephone is utilized here as an example to illustrate the implementation of the NSGA-II algorithm. The chromosome will be codified in the form of a string consisting of 15 genes, representing 2 assemblies and 13 subcomponents, and each gene has a value from 1 to 5. The initial population consists of 1000 random chromosomes that satisfy the constraints prescribed in Equations 3.8–3.10. For example, if Assembly #1 is assigned to be recycled, all of its subcomponents 2–5 should remain within Assembly #1. Once the population has been initialized, they will be sorted based on non-domination into different fronts using the economic and environmental objectives stated in Equations 3.6 and 3.7. Based on the sorting results, the best population will be selected to generate the offspring population, using the uniform crossover and multiple point mutations. After that, the population with the current population and current offspring will be sorted again. To achieve an efficient convergence to a high-quality Pareto curve, the algorithm is executed through 200 loops. This technique has proven to be very effective in finding a wide spread of economic and environmental Pareto solutions for both cases, as shown in Figures 3.4 and 3.5 and Tables 3.3 and 3.4.

3.1.4 RESULTS AND DISCUSSION

Steps I and II of the decision support tool provide a systematic process to evaluate the feasibility of a product and its components for remanufacturing. As described earlier, the evaluation process comprises a sequential examination of characteristics

TABLE 3.2

Economic and Environmental Data for Business IP Telephone

Name	Eco_Reman ($)	Eco_Shrd/Rc ($)	Eco_Lf ($)	Eco_Dis ($)	Evn_Reman (kJ)	Evn_Shrd/Rc (kJ)
AS1	$-\infty$	$-\infty$	-3.90	-1.99	0.00	0.00
C2	-0.20	-0.76	-0.76	$-\infty$	$-\infty$	1860.74
C3	-0.20	-0.76	-0.76	$-\infty$	6119.65	1751.41
C4	-0.74	-0.77	-0.80	$-\infty$	1432.90	1130.85
C5	-0.47	-0.39	-0.39	$-\infty$	655.55	398.64
C6	-0.11	$-\infty$	-0.16	$-\infty$	566.42	0.00
C7	-0.42	$-\infty$	-0.22	$-\infty$	96.49	0.00
C8	-1.04	-0.22	-0.24	$-\infty$	4544.81	3007.55
C9	-1.08	-0.26	-0.27	$-\infty$	2897.78	1916.82
AS10	$-\infty$	$-\infty$	-29.35	-0.66	0.00	0.00
C11	-0.18	-2.57	-2.56	$-\infty$	18,731.10	5360.74
C12	-0.08	-0.14	-0.14	$-\infty$	538.39	286.79
C13	-0.44	-0.41	-0.42	$-\infty$	188.04	139.91
C14	-0.56	-0.98	-0.99	$-\infty$	6045.80	4590.42
C15	-0.20	-0.20	-0.20	$-\infty$	1986.50	1817.75
C16	-1.14	$-\infty$	-3.92	$-\infty$	742.05	0.00
C17	-1.24	$-\infty$	-5.39	$-\infty$	1765.27	0.00
C18	-4.06	-4.35	-4.35	$-\infty$	4037.28	3186.39
C19	-0.20	-0.45	-0.45	$-\infty$	2239.01	1493.59
C20	-0.20	-0.51	-0.51	$-\infty$	1253.98	836.50
C21	-0.20	-0.55	-0.55	$-\infty$	1121.98	748.44
C22	-0.23	-0.07	-0.07	$-\infty$	43.89	12.56
C23	-0.23	-0.22	-0.22	$-\infty$	1456.77	867.36
C24	-0.23	-0.07	-0.07	$-\infty$	357.58	212.90
C25	-0.23	-0.07	-0.07	$-\infty$	131.67	37.68
C26	-0.23	-0.07	-0.07	$-\infty$	87.77	25.12
C27	-0.23	-0.27	-0.27	$-\infty$	87.77	25.12
C28	-0.23	-0.40	-0.40	$-\infty$	570.54	163.28
C29	-0.23	-0.22	-0.22	$-\infty$	781.20	223.57
C30	-0.23	-0.09	-0.09	$-\infty$	87.77	25.12
C31	-0.16	-0.23	-0.23	$-\infty$	623.32	367.81
C32	-0.20	-0.48	-0.48	$-\infty$	465.02	274.40
C33	-1.23	$-\infty$	$-\infty$	$-\infty$	392.77	0.00
C34	-0.24	-0.70	-0.70	$-\infty$	359.88	102.99
C35	-0.07	-0.52	-0.52	$-\infty$	105.30	64.05
C36	-0.18	-1.25	-1.26	$-\infty$	26,701.04	7641.69
C37	-0.15	$-\infty$	$-\infty$	$-\infty$	379.50	0.00
C38	-0.18	-0.90	-0.91	$-\infty$	12,016.35	3439.01
C39	-0.15	$-\infty$	$-\infty$	$-\infty$	379.50	0.00

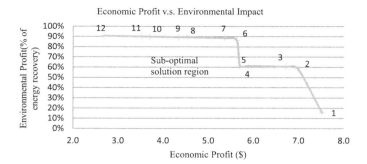

FIGURE 3.4   Optimal EOL strategy set for consumer telephone.

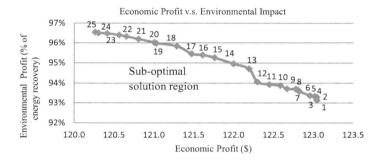

FIGURE 3.5   Part of optimal EOL strategy set for business IP telephone.

TABLE 3.3
Optimum EOL Strategy Set for Consumer Telephone

							Comp								
Solution	AS 1	C2	C3	C4	C5	C6	C7	AS8	C9	C10	C11	C12	C13	C14	C15
1	4	1	1	2	1	2	2	3	5	5	5	5	5	5	5
2	4	1	1	2	1	2	2	4	1	2	1	1	3	1	1
3	4	1	1	2	1	2	2	4	1	2	1	1	1	1	1
4	4	1	1	2	1	2	1	4	1	2	1	1	1	1	1
5	4	1	1	1	1	2	2	4	1	2	1	1	1	1	1
6	4	1	1	2	1	2	2	4	1	1	1	1	3	1	1
7	4	1	1	2	1	2	2	4	1	1	1	1	1	1	1
8	4	1	1	2	1	1	2	4	1	1	1	1	1	1	1
9	4	1	1	1	1	2	2	4	1	1	1	1	1	1	1
10	4	1	1	2	1	1	1	4	1	1	1	1	1	1	1
11	4	1	1	1	1	1	2	4	1	1	1	1	1	1	1
12	4	1	1	1	1	1	1	4	1	1	1	1	1	1	1

1: remanufacturing; 2: shredding/recycling; 3: landfill; 4: disassemble; 5: remain within assembly.

TABLE 3.4
Part of Optimum EOL Strategy Set for Business IP Telephone

Comp

Sol	AS1	C2	C3	C4	C5	C6	C7	C8	C9	AS10	C11	C12	C13	C14	C15	C16	C17	C18	C19	C20	C21	C22	C23	C24	C25	C26	C27	C28	C29	C30	C31	C32	C33	C34	C35	C36	C37	C38	C39
1	4	1	1	2	1	3	2	2	2	4	1	1	2	1	2	1	1	1	1	1	1	2	1	2	2	2	1	1	1	2	1	3	3	1	1	1	1	1	1
2	4	1	1	2	1	3	2	2	2	4	1	1	2	1	1	1	1	1	1	1	1	2	1	2	2	2	1	1	1	2	1	3	3	1	1	1	1	1	1
3	4	1	1	2	1	3	2	2	2	4	1	1	1	1	1	1	1	1	1	1	1	2	1	2	2	2	1	1	1	2	1	3	3	1	1	1	1	1	1
4	4	1	1	2	1	3	2	2	2	4	1	1	2	1	2	1	1	1	1	1	1	2	1	2	2	2	1	1	1	2	1	3	3	1	1	1	1	1	1
5	4	1	1	2	1	3	2	2	2	4	1	1	2	1	1	1	1	1	1	1	1	2	1	2	2	2	1	1	1	2	1	3	3	1	1	1	1	1	1
6	4	1	1	2	1	3	2	2	2	4	1	1	2	1	1	1	1	1	1	1	1	2	1	2	2	2	1	1	1	2	1	3	3	1	1	1	1	1	1
7	4	1	1	2	1	3	2	2	2	4	1	1	2	1	1	1	1	1	1	1	1	2	1	2	2	2	1	1	1	2	1	3	3	1	1	1	1	1	1
8	4	1	1	2	1	3	2	2	2	4	1	1	1	1	1	1	1	1	1	1	1	2	1	2	2	2	1	1	1	2	1	3	3	1	1	1	1	1	1
9	4	1	1	2	1	3	2	2	2	4	1	1	2	1	1	1	1	1	1	1	1	2	1	2	2	2	1	1	1	2	1	3	3	1	1	1	1	1	1
10	4	1	1	2	1	3	2	2	2	4	1	1	2	1	1	1	1	1	1	1	1	2	1	2	2	2	1	1	1	2	1	1	1	1	1	1	1	1	1
11	4	1	1	2	1	3	2	2	2	4	1	1	1	1	1	1	1	1	1	1	1	2	1	2	2	2	1	1	1	1	1	1	1	1	1	1	1	1	1
12	4	1	1	2	1	3	1	2	2	4	1	1	2	1	1	1	1	1	1	1	1	2	1	2	2	2	1	1	1	2	1	3	3	1	1	1	1	1	1
13	4	1	1	2	1	3	1	2	2	4	1	1	2	1	1	1	1	1	1	1	1	2	1	2	2	2	1	1	1	2	1	3	3	1	1	1	1	1	1
14	4	1	1	2	1	3	1	2	2	4	1	1	2	1	1	1	1	1	1	1	1	2	1	2	2	2	1	1	1	2	1	1	1	1	1	1	1	1	1
15	4	1	1	2	1	3	1	2	2	4	1	1	2	1	1	1	1	1	1	1	1	2	1	2	2	2	1	1	1	2	1	1	1	1	1	1	1	1	1
16	4	1	1	2	1	3	1	2	2	4	1	1	2	1	1	1	1	1	1	1	1	2	1	2	2	2	1	1	1	2	1	3	3	1	1	1	1	1	1
17	4	1	1	2	1	3	1	2	2	4	1	1	1	1	1	1	1	1	1	1	1	2	1	2	2	2	1	1	1	2	1	3	3	1	1	1	1	1	1
18	4	1	1	2	1	3	1	2	2	4	1	1	2	1	1	1	1	1	1	1	1	2	1	2	2	2	1	1	1	2	1	3	3	1	1	1	1	1	1
19	4	1	1	2	1	3	1	2	2	4	1	1	2	1	1	1	1	1	1	1	1	2	1	2	2	2	1	1	1	1	1	1	1	1	1	1	1	1	1
20	4	1	1	1	1	1	1	1	1	4	1	1	2	1	1	1	1	1	1	1	1	2	1	2	2	2	1	1	1	2	1	1	1	1	1	1	1	1	1
21	4	1	1	1	1	1	1	1	1	4	1	1	2	1	1	1	1	1	1	1	1	2	1	2	2	2	1	1	1	1	1	1	1	1	1	1	1	1	1
22	4	1	1	1	1	1	1	1	1	4	1	1	2	1	1	1	1	1	1	1	1	2	1	2	1	2	1	1	1	2	1	1	1	1	1	1	1	1	1
23	4	1	1	1	1	1	1	1	1	4	1	1	2	1	1	1	1	1	1	1	1	2	1	1	1	2	1	1	1	1	1	1	1	1	1	1	1	1	1
24	4	1	1	1	1	1	1	1	1	4	1	1	1	1	1	1	1	1	1	1	1	2	1	2	1	2	1	1	1	2	1	1	1	1	1	1	1	1	1
25	4	1	1	1	1	1	1	1	1	4	1	1	1	1	1	1	1	1	1	1	1	2	1	1	1	1	1	1	1	2	1	1	1	1	1	1	1	1	1

1: remanufacturing; 2: shredding/recycling; 3: landfill; 4: disassemble; 5: remain within assembly.

that define the prerequisites for carrying out product or part remanufacturing. Failing to fulfill one or more of the prerequisites will mean that the external conditions would be working against the product or part remanufacturing efforts. In addition, the sequential examination process aims to stimulate process modification or product redesign in the case where a product or a component is not feasible for remanufacturing. For example, if technical data required to restore a product is not available while the product has been returned for remanufacturing, reverse engineering can be carried out to extract the needed technical data. Another example is to improve the design viability of a product or subassembly through decoupling the design elements that will suffer early obsolescence from the more stable product systems, and thus extend the product valuable lifetime.

In the third step, the EOL strategy planning is optimized with respect to two objectives, namely, maximizing economic benefit and minimizing environmental impact. As shown in Figures 3.4 and 3.5, each point on the Pareto curve represents an optimal EOL strategy between cost and environmental actions. Moving along the Pareto line from the rightmost strategy to the leftmost strategy, the economic return decreases with an increase in energy recovery rate. Using Figure 3.4 as an example, Strategy #1 on the curve is the maximum profit strategy with the lowest energy recovery rate, which involves complete disassembly of Assembly #1 and landfilling Assembly #8 as a whole unit; Strategy #12, located on the other extreme, corresponds to the maximum energy recovery strategy with the lowest economic return, which suggests complete disassembly of the product and remanufacturing of every single component. The business IP telephone remanufacturing has shown both economic and environmental advantages over the consumer telephone remanufacturing; at the maximum profit strategy, 86% of the embodied energy of the business telephone can be preserved, whereas for the consumer telephone, only 15% of the product energy can be retained.

It can be noted that there are two abrupt kinks on the Pareto curve of Figure 3.4, which involve 43% and 25% increase in the energy recovery rate at Strategy #2 and Strategy #6. On a closer examination, some redesign ideas may be inspired, such as incorporating design elements to facilitate the disassembly of Assembly #8, or redesigning component #10 with elements to facilitate its upgrade, replacement or other forms of enhancement. These redesign ideas can be tested with the optimization model to examine their impact on EOL recovery strategy. For example, as shown in Figure 3.6, the shape of the Pareto curve has changed with a decrease in the disassembly cost of AS8. The "tipping scenario" happens when the disassembly cost of AS8 decreases to below 50%. In this scenario, the EOL strategy of AS8 will change from landfill to disassembly with all its components remanufactured, recycled or disposed of.

The EOL strategy planning process can be rather complex and dynamic in the real world. Factors, such as the value and cost of components, replacement parts availability, technology availability, are not stable and will affect the EOL decision making to various degrees. To address the complexity of EOL decision making, sensitivity analysis with respect to different situational variables can be carried out to understand their impact on EOL Pareto solutions.

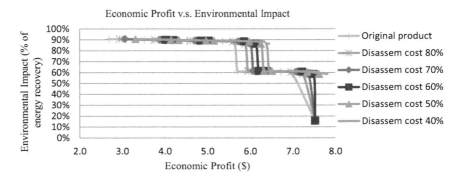

FIGURE 3.6 Impact of design change of consumer telephone on Pareto set.

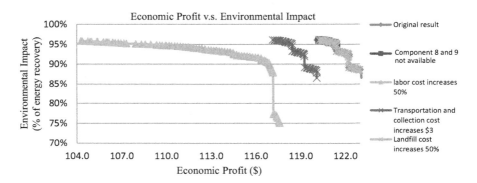

FIGURE 3.7 Multi-situational EOL strategy graph for business IP telephone.

Figure 3.7 illustrates the multi-situational strategy graphs constructed for the business IP phone case study. The trade-off sets are provided for the base case along with four other scenarios featuring differences in replacement part availability, labor cost, transportation and collection cost, landfill cost. The results have provided significant insights regarding the impact of these situational factors on EOL decisions. For example, the addition of the transportation and collection cost will cause a parallel shift of the Pareto curve, as the incurred cost is applied on the product level. The change of labor cost will affect the remanufacturing cost of each component and thus incur the noticeable change of the shape of the Pareto curve. The replacement part availability and remanufacturing technology availability can affect the EOL choices and thus the overall remanufacturing strategies. For example, when replacement parts for C8 and C9 are not available, the returned parts will need to be remanufactured. Comparing with other scenarios, the Pareto curve is less affected by the change of landfill cost in this case study, as most of the components are planned to be remanufactured or shredded/recycled.

The applicability of the proposed methodology has been demonstrated using relatively simple desktop phones. However, it can handle products with a large number of components due to the efficiency and effectiveness of NSGA-II to identify optimal EOL solutions in a large search space. The foreseeable limitation may come from data acquisitions for the input of this methodology, which includes the remanufacturing cost and the replacement cost of each component, and the disassembly and reassembly costs. This limitation can be addressed by relating to similar product EOL information or consulting with remanufacturing experts for a reasonable estimation. Other information, such as material recycling value, landfill cost, embodied energy, secondary production energy, can be stored in the database. Once the product information, such as Bill of Materials, has been keyed in, the economic and environmental profit can be calculated automatically, to reduce the burden of data acquisition. Meanwhile, it is noted that the economic profit and environmental impact are the two objectives optimized in this study. To capture other dimensions of considerations, such as production cost, product performance, additional metrics can be added to the objective functions as the NSGA-based methodology is general in dimensionality and can be utilized for establishing Pareto sets, which consider trade-offs between more than two metrics simultaneously. Therefore, the proposed methodology can serve to accelerate the diffusion of remanufacturing requirements into original product design requirements by providing an approach that decision makers can use to quantify and visualize the trade-offs between requirements from different disciplines.

In addition, the proposed optimization model deals with the economic performance of one product type in a decoupled way, i.e., without considering the financial synergy resulted from sharing reverse logistic cost or setup cost with other product types within the factory. When a full-scale treatment of economic analysis is carried out, the proposed methodology can be iterated with the readjusted cost value, such as the reverse logistic cost or resale price, to recalculate the EOL strategy on the single-product level.

3.1.5 SUMMARY

A decision support tool to facilitate product remanufacturing strategy planning is presented in this section. This methodology has advanced the previous research in EOL decision planning in the following three features: First, the proposed methodology provides a holistic approach, where a four-step decision tool will guide decision makers toward optimum EOL strategies decisions, through addressing the comprehensive aspects of remanufacturing considerations; second, the methodology provides flexibility by utilizing NSGA-II to determine explicitly a Pareto set of optimal EOL solutions to facilitate the effort of the decision makers to maximize the environmental benefit of remanufacturing for a given economic profit. Lastly, the methodology is comprehensive as the Pareto sets of optimum solutions can be calculated within a reasonable computational time, which permits extensive sensitivity analysis to understand the impact of situational variables on EOL decision making thoroughly, and thus leading to the improved EOL decision making and product design. The applicability of NSGA-II for determining the Pareto set of optimal EOL solutions has been demonstrated numerically with two desktop phones case studies.

3.2 DESIGN FOR REMANUFACTURING

3.2.1 INTRODUCTION

Besides assessing the remanufacturability of products and components at the early design stage, the proposed solution will take a proactive measure to improve the potential of a product for remanufacturing through developing an effective and efficient product design support tool. As 80% of the cost of the product is committed at the design stage, it is of significance to address the remanufacturing issue and concern during this stage (David and Anderson, 2014). Previous research has presented comprehensive design for remanufacturing guidelines to address and mitigate the difficulties involved in each remanufacturing step, like design for disassembly, design for cleaning, design for reconditioning, etc. To be able to consider each remanufacturing aspect individually could be the most effective method. However, in reality, this may be an overly daunting and time-consuming task for designers. Most of the design guidelines for remanufacturing are fairly general and rarely consider how these design guidelines may fit in with the already-sophisticated design process (Hatcher et al., 2011). Meanwhile, it is a widely recognized belief that design for remanufacturing (DfRem) is most effective when implemented in the early design stage, as few decisions have been made and the design freedom is large (Amezquita et al., 1995; Zwolinski et al., 2006). However, many of the DfRem tools being proposed by academia, especially those of quantitative nature, require too much technical data. Thus, these tools are either too complex to be used at the early design stage, or by the time product specification has been defined, are too late to make substantial changes to the design (Hatcher et al., 2011). In addition, design for remanufacturing should not be considered in an isolated manner. Given the potential conflict that DfRem may have with other Design for X (DfX) methodology, such as assembly and manufacture, there is a need for an analysis that can demonstrate properly the extent of the impact DfRem has on the remanufacturing process and other life cycle stages that are involved (Zwolinski et al., 2006).

To address the above-mentioned limitations, a holistic decision support tool for DfRem is developed and presented in this section. This approach will steer a product design toward higher remanufacturability from four major design aspects, namely, material selection, material joining methods, structure design and surface coating, to address the various aspects of remanufacturing concerns. While these four aspects are only a subset of product design considerations, they are selected because they are particularly relevant to the realization of remanufacturing objectives. The design for remanufacturing requirements or criteria will be presented first in a manner that designers are familiar with to reduce the complexity of the design process. After that, a decision support tool based on the MCDM technique, namely, Fuzzy Technique of ranking Preferences by Similarity to the Ideal Solution (Fuzzy TOPSIS), will be presented to evaluate the impact of DfRem on remanufacturing efficiency. The selected design alternatives will be further compared in a proposed multiple life cycle assessment model to examine the impact of remanufacturability enhancement design features on the overall life cycle performance, so as to improve the effectiveness and robustness of the decision change. In addition, to assist decision makers in the application of the proposed methodology and to simplify computation complexity, a computation tool based on the Visual C# has been developed, which allows for fast computation and ease of use of this methodology. The applicability of the decision support tool will be demonstrated using automotive parts design.

3.2.2 Major DfRem Considerations

Materials selection is at the core of decision making throughout product design development, as the material properties can influence the various aspects of a product life cycle, such as manufacturing cost, market acceptance and functional performance (Andrea and Brown, 1993). The product remanufacturing performance, including component recoverability, economic incentive, and the amount and toxicity of the waste generated through remanufacturing, is also largely influenced by the properties of the materials used (Charter and Gray, 2008). To facilitate the remanufacturing process, it is desirable for the materials of a product to be durable so as to enhance the service life of the components and prevent the core from breaking down during remanufacturing. It is desirable that properties of the materials are adaptable to cleaning and reconditioning process.

A critical design aspect that influences product fit, form and function is the material joining method. Typically, material joining involves utilizing various methods to affix two or more objects together, e.g., bolts, nuts, screws, rivets, staples, magnets, retaining rings, adhesive joints, welding and crimping. It is an essential factor that has to be taken into account for EOL consideration, as the way the parts are joined together can facilitate or impede product disassembly for reuse, remanufacturing and recycling (Amezquita et al., 1995; Ijomah et al., 2007; Sundin and Lindahl, 2008). When products are disassembled for remanufacturing, there are usually three types of scenarios planned for its joints/fasteners and adjoining parts. The first scenario is disassembly without destruction, including joints/fasteners. In this scenario, adjoining parts are intended for reuse and/or remanufacturing and the condition of the joints/fasteners after disassembly is important. The second scenario is disassembly without destruction, excluding joints/fasteners. In this situation, disassembly without any degradation to adjoining parts is desired so that they can be reused for remanufacturing. However, the condition of the joints/fasteners after disassembly is not critical and thus the joints/fasteners are allowed to be destructed if necessary. The third scenario is disassembly with allowable destruction. This would be in a recycling context, where the separation of the parts is important, yet damage to certain parts and joints/fasteners is acceptable. Given the three different EOL scenarios, it is critical that the decision makers prioritize the EOL scenarios and rate the performance of the candidate joining methods toward each scenario accordingly. Besides, the cost and the environmental consideration should be accounted during the joining methods selection process.

Another design aspect that is closely related to product remanufacturing efficiency is the product structure design. Product remanufacturing, especially in complex product remanufacturing, is a challenging task as it involves the disassembly process to separate different materials and retrieve the reusable components in a non-destructive and cost-effective manner, which is closely related to the way the components are arranged and interacted upon in the product (Kuo, 2006; Chu et al., 2009).

Meanwhile, the number, design tolerance, shape and position of components will also affect the efficiency of various remanufacturing processes, such as cleaning, inspection, reconditioning. For example, if a part to be replaced is located deeply inside a product, accessing and retrieving the part becomes challenging and could increase the cost of remanufacturing. If the number and types of components of a product is large, it will increase the complexity of components discerning and classification as well as the possibility of selecting wrong components when performing reassembly.

Surface coating is a critical aspect that influences the potential of a product for remanufacturing. Usually, when a substrate material has been chosen for its bulk design characteristics and it may not possess the desirable surface properties, surface coating will be applied to the substrate to meet certain surface requirements, such as surface fatigue resistance, wear, corrosion, or for aesthetic purposes. Improper selection of surface coating methods can increase the failure frequency of product caused by material wear or corrosion and add burden to product remanufacturing process substantially, e.g., a very smooth surface coating may involve substantial effort to be restored to a like-new condition, or a texture that is too coarse may trap dirt easily and complicate the cleaning process (Sundin and Lindahl, 2008). Thus, the selection of the surface coating should account for the environment in which a component will face and the degradation factors that may cause component failure, so as to enhance the product durability performance while facilitate the cleaning and restoration process that the core will go through.

The detailed remanufacturing evaluation criteria with respect to these four design aspects are compiled and presented in Table 3.5. It is noted that the design considerations, presented in Table 3.5, are meant to be applied to the components or parts that have been identified to have the potential for remanufacturing, especially those components that have high embedded value, long technology life cycle or high durability, e.g., engine, turbocharger, starter and alternator. It may not be sensible to be applied to the situations where there is an industry standard for certain design requirement (Yang et al., 2016).

3.2.3 EVALUATING THE IMPACT OF DFREM ON REMANUFACTURING PERFORMANCE

The proposed criteria and consideration can be used either as a guideline to assist the product designers at the early stage or as a set of criteria for evaluating the impact of alternative design features on product remanufacturability. An optimal selection from a finite set of feasible alternatives and a predetermined number of criteria is often regarded as a MCDM problem. Among these MCDM methods, Fuzzy TOPSIS is chosen in this research to evaluate the design concepts with respect to their performance in the remanufacturing process and aim to enable designers to make better design choices and enhance the opportunity for product remanufacturing. The performance of different design concepts or choices, like material selection, material joining method, structure design and surface coating, will be evaluated based on the remanufacturing considerations, including durability, cleanability, restorability and upgradability, EHS, cost and complexity, as presented in Table 3.5. This is a research area that has not been explored previously but deserves attention due to

TABLE 3.5

Evaluation Criteria from Remanufacturing Perspective

	Material	Material Joining Method	Structure Design	Functional and Decorative Surface Coating
Durability	• Corrosion resistance • Wear resistance • Fatigue resistance	• Corrosion resistance		• Wear/Corrosion/Surface fatigue resistance (Functional coating) • Fingerprint/Scratch resistance (Decorative coating) • Adhesion
Disassemblablity and assemblability		• Disassembly without destruction, (include fastener/joint) • Disassembly without destruction, (exclude fastener/joint) • Disassembly, destruction allowed (for recycling) • Ease of reassembly	• Modularity for easy separation • Accessibility to valuable and reusable components	
Cleanability	• Ease of removing impurity and deposit • Resistance to cleaning	• Ease of removing impurity and deposit	• Avoid intricate or unnecessary concealed design form	• Ease of removing the contaminants (coating removal is not required) • Potential damage to the substrate (coating removal is required)
Restorability/upgradability	• Ease of receiving machining process • Ease of receiving additive process • Ease of receiving conditioning process • Reliability of the reconditioned part	• Standardization of joining method	• Accessibility to the failure prone parts • Tolerance design for multiple life cycle • Modularity for replacement/upgradability	• Ease of receiving surfacing engineering

(Continued)

TABLE 3.5 (*Continued*)
Evaluation Criteria from Remanufacturing Perspective

	Material	Material Joining Method	Structure Design	Functional and Decorative Surface Coating
Environmental Health and Safety (EHS)	• Recyclability • Air emissions and waste disposal • Toxicity • Scarcity of raw material • Law and regulation	• Compatibility with other parts • Toxicity	• Recyclability	• Air emissions and waste disposal • Recyclability • Law and regulation
Cost	• Raw material cost	• Labor cost • Capital cost	• Labor cost • Capital cost	• Labor cost • Material and energy consumption • Capital cost
Complexity	• No. of material	• No. of types of fastener/joint • No. of fastener/joint • Tool standardization • Accessibility to fastener/joint	• No. of parts and components • Standardization of parts and components	• Compatibility with substrate material

the environmental and economic benefits that can be achieved through automotive remanufacturing as well as the impact of initial product design on product remanufacturability. The framework of this decision support tool is presented next.

3.2.3.1 Step 1: Determining the Candidate Components for Remanufacturing

The proposed decision support system is meant to be applied to the components or parts that have been identified to have the potential for remanufacturing. However, during the remanufacturing process, it is unlikely that all the components of the returned products will be remanufactured. The detailed guidelines for identifying the candidate product and its components for remanufacturing can be found in Section 3.1.

3.2.3.2 Step 2: Selecting the Evaluation Aspect and Design Candidates

The set of remanufacturing evaluation metrics with respect to material selection, material joining method, structure design and surface coating are compiled and shown in Table 3.5. Decision makers will need to select the design aspect they want to examine and select the feasible design candidates for evaluation and comparison.

3.2.3.3 Step 3: Building Performance Matrix

The performance of candidate materials against the evaluation criteria is collected from the SAMSG (Sustainable Automotive Materials Selection Guide) (NCMS, 2012), the material handbooks (Bauccio, 1993; Davis, 1996; Murray, 1997), website information (Guesser et al., 2004; Dawson and Indra, 2007), as well as remanufacturing experts, who have over 25 years of remanufacturing working/research experience. The variable r_{ij} is used to represent the performance rating of ith material alternative with respect to jth evaluation criterion, and the material performance matrix R is thus expressed as Equation 3.11, where $X_1 \dots X_m$ are the design candidates and $A_1 \dots A_n$ are the evaluation criteria.

$$
\begin{array}{c}
X_1 \dots X_i \dots X_m \\[4pt]
R = \begin{array}{c} A_1 \\ \vdots \\ A_j \\ \vdots \\ A_n \end{array}
\begin{bmatrix}
r_{11} & \cdots & r_{i1} & \cdots & r_{m1} \\
\vdots & & \vdots & & \vdots \\
r_{1j} & \cdots & r_{ij} & \cdots & r_{mj} \\
\vdots & & \vdots & & \vdots \\
r_{1n} & \cdots & r_{in} & \cdots & r_{mn}
\end{bmatrix}
\end{array} \tag{3.11}
$$

3.2.3.4 Step 4: Calculating the Weight Factor

As a product or component will be used in different application environments by different end users, and has different design restrictions, the design evaluation criteria are usually not of equal importance relative to each other. Therefore, some forms of weighting shall be introduced as part of the evaluation process. To obtain a more reasonable weight coefficient, the weight w_j for the jth criterion will be a combination of two sets of weights, as shown in Equation 3.12, where α_j is the weight obtained via the entropy method (Shannon and Weaver, 1947) and β_j is the subjective weight assigned by experts from the remanufacturing field. In Equation 3.12, j is the number of the criterion and n is the total number of criteria.

$$w_j = \frac{\alpha_j * \beta_j}{\sum\limits_{j=1}^{n} \alpha_j * \beta_j} \quad j = 1,\ldots,n \tag{3.12}$$

a. The set of weights from the entropy method

The entropy method makes use of the information that is already contained in the defined material performance matrix R and uses the probability theory to derive directly the relative importance of the evaluation criteria. The underlying principle is to assess the uncertainty in the information, as there is a common agreement that a broader distribution represents a greater uncertainty than that of a sharply peaked one (Shannon and Weaver, 1947). This method consists of the following procedures.

1. Normalization of the decision matrix R

$$P_{ij} = \frac{r_{ij}}{\sum\limits_{i=1}^{m} r_{ij}} \quad i = 1,2,\ldots, m; \quad j = 1, 2,\ldots,n; \tag{3.13}$$

In Equation 3.13, i is the number of the alternative, m is the total number of alternatives and P_{ij} represents the normalized performance matrix.

2. Calculate the Entropy E_j of the normalized values of jth criterion

$$E_j = -\left(\frac{1}{\log(m)}\right)\sum\limits_{i=1}^{m} P_{ij} \log(P_{ij}) \quad i = 1,2,\ldots, m; \quad j = 1, 2,\ldots,n; \tag{3.14}$$

The calculated value of the Entropy E_j will be in the range of [0–1].

3. Calculate the weight α_j of the Entropy of jth criterion

$$\alpha_j = \frac{|1 - E_j|}{\sum\limits_{j=1}^{n} |1 - E_j|} \quad j = 1, 2,\ldots,n; \tag{3.15}$$

If P_{ij} for jth criterion has wide range, it will yield a small value of E_j, which will result in the large weight factor α_j.

b. Subjective weight from expert's input

The subjective weights are assigned by experts from the remanufacturing field based on their professional judgment and past experience; it is a

TABLE 3.6
Crisp Value for Subjective Importance Rating

Rating of Relative Importance	Very Low (VL)	Low (L)	Medium (M)	High (H)	Very High (VH)
Crisp value	1	3	5	7	9

simplified way of weight determination and reduces the computational time and complexity. A numerical approximation system is used to convert the linguistic judgment systematically to their corresponding crisp score, as shown in Table 3.6.

3.2.3.5 Step 5: Developing Ranking for Design Candidates Using the Fuzzy TOPSIS Method

Despite its effectiveness in concept, TOPSIS is often criticized for its inability to deal with the vagueness and uncertainty involved in the judgment process. Hence, the fuzzy set theory is proposed to be combined with the TOPSIS method, which is also known as the Fuzzy TOPSIS approach. This combined approach can handle the imprecise information by converting them into linguistic variables, which will be expressed using a triangular fuzzy number, i.e., $\left[a_{ij}, m_{ij}, b_{ij} \right]$, as illustrated in Table 3.7. In this way, a higher degree of uncertainty can be included in the decision-making process.

According to the concept of TOPSIS, the relative closeness value C_i^+ is introduced to determine the ranking order of the alternatives, by calculating the distance of the alternative to the ideal solution, namely, S_i^+ and its distance to the non-ideal solution, namely, S_i^-. The larger the closeness value, the better is the design alternative. In the fuzzy environment, the distance between two triangular fuzzy numbers will be calculated using a vortex method. Detailed calculation steps for carrying out Fuzzy TOPSIS can be referred to Chen's work (2000) and Wang and Chan's work (2013a).

3.2.3.6 Step 6: Compatibility Check

Once the ranking for the design candidates for the designed part has been obtained, decision makers may proceed with product internal check to detect whether there are any potential conflicts with the design choices or design constraints of other parts, such as different coefficient of thermal expansion between the neighboring components or decrease in functional performance. If the selected design candidate for the designed component may lead to constraints in the options available for other

TABLE 3.7
Linguistic Rating r_{ij} and Its Corresponding Triangular Fuzzy Number

Rating of Design Performance	Poor (P)	Medium Poor (MP)	Fair (F)	Medium Good (MG)	Good (G)
Fuzzy number	[0,1,3]	[1,3,5]	[3,5,7]	[5,7,9]	[7,9,10]

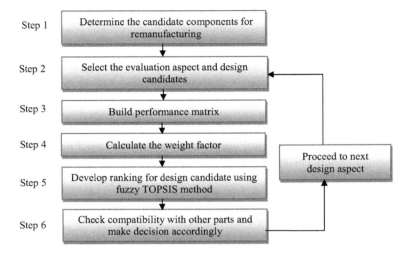

FIGURE 3.8   Flow chart for the proposed decision support tool.

elements, the next optimum design candidate will be chosen. A flow chart of the proposed methodology for designing automotive products for remanufacturing is shown in Figure 3.8 (Yang et al., 2017).

3.2.4   EVALUATING THE IOF DFREM ON PRODUCT LIFE CYCLES

Product design for remanufacturing cannot be viewed in an isolated manner. As mentioned by several researchers (Shu and Flowers, 1999; Ijomah et al., 2007), DfRem is often in conflict with other DfX methodology, such as manufacturing and environment. An approach that can assess the impact of remanufacturability enhancement features on the overall product life cycle and deliver a robust and comprehensive remanufacturing design suggestion is of great importance. To address this need, a decision support tool is proposed that incorporates the "life cycle thinking" to evaluate the economic and environmental impact of remanufacturing enhancement features over multiple usage cycles. Meanwhile, situational variables, such as successful remanufacturing rate and the number of life cycles, will be accounted for in the model to examine their impact on the overall life cycle performance.

The traditional "cradle-to-grave" product life cycle has been adapted to "cradle-to-cradle", as graphically represented in Figure 3.9. Basically, six generic phases,

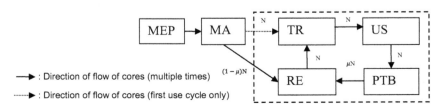

FIGURE 3.9   Flow of products over multiple life cycles.

namely, MEP, manufacturing (MA), transportation (TR), usage stage (US), product take-back (PTB), and remanufacturing (RE), have been included to describe the life cycle of a product. The method assumes that the number of products within a system is N, and the first usage cycle will be composed of the newly manufactured products only. During the remanufacturing stage, X_{ij} cores will be reprocessed to "good as new" quality to enter into the next life cycle, while the lost cores will be made up with $(1 - \mu) * N$ virgin products.

Cumulated energy demand (CED) will be used to represent the sum of the primary energy demand throughout the life span of a product and thus approximates the holistic environmental performance of remanufacturing enhancement features. To apply the concept of CED to the above-described life cycle model, CED will be calculated as a function of the number of usage cycles M, the successful remanufacturing rate μ and the primary energy demand for MEP CED_{MEP}, manufacturing CED_{MA}, transportation CED_{TR}, usage CED_{US}, PTB CED_{PTB} and remanufacturing CED_{RE} of the product, as shown in Equation 3.16.

Similarly, cumulated cost (CC) will be used to represent the sum of the expense throughout the life span of a product and its mathematical expression is shown in Equation 3.17.

$$CED_{Total} = CED_{MEP_1} + CED_{MA_1} + \sum_{i=1}^{M} \left(CED_{US_i} + CED_{TR_i} + CED_{PTB_i} \right)$$

$$+ \sum_{i=1}^{M} \left(\mu_i * CED_{RE_i} + (1 - \mu_i) * \left(CED_{MA_i} + CED_{MEP_i} \right) \right) \qquad (3.16)$$

$$CC_{Total} = CC_{MEP_1} + CC_{MA_1} + \sum_{i=1}^{M} \left(CC_{US_i} + CC_{TR_i} + CC_{PTB_i} \right)$$

$$+ \sum_{i=2}^{M} \left(\mu_i * CC_{RE_i} + (1 - \mu_i) * \left(CC_{MA_i} + CC_{MEP_i} \right) \right) \qquad (3.17)$$

where:

M: number of life cycles;
μ: successful remanufacturing rate;
i: ith life cycle;
CED_{Total}: CED throughout the entire product life span;
CC_{Total}: CC throughout the entire product life span;
CED_{MEP_i}: CED for the MEP during the ith life cycle;
CC_{MEP_i}: CC for the MEP during the ith life cycle;
CED_{MA_i}: CED for the product manufacturing during the ith life cycle;
CC_{MA_i}: CC for the product manufacturing during the ith life cycle;
CED_{TR_i}: CED for the transportation of product during the ith life cycle;
CC_{TR_i}: CC for the transportation of product during the ith life cycle;

CED_{US_i}: CED for the use of product during the ith life cycle;
CC_{US_i}: CC for the use of product during the ith life cycle;
CED_{PTB_i}: CED for the take-back of product during the ith life cycle;
CC_{PTB_i}: CC for the take-back of product during the ith life cycle;
CED_{RE_i}: CED for the product remanufacturing during the ith life cycle;
CC_{RE_i}: CC for the product remanufacturing during the ith life cycle.

It should be noted that for a complex life cycle that involves product remanufacturing, designers have to estimate the maximum number of life cycles M that a product or component can support. During the remanufacturing process, it is unlikely that 100% of the product can be remanufactured successfully to "good as new" quality due to recovery process capability (poor quality of returned cores, technology constrains, low recoverable value, etc.). Hence, an estimation of the successful remanufacturing rate is required. Further, only the remanufacturing strategy is considered as the closed-loop strategy, as it is the main focus of this study and is able to preserve more significant amount of embedded energy of the product than other closed-loop strategies. The inclusion of other recovery strategies, such as recycling or disposal, will be addressed in the future study.

3.2.5 OVERALL APPROACH FOR PRODUCT DESIGN FOR REMANUFACTURING AND ITS SOFTWARE IMPLEMENTATION

The Design for Remanufacturing and Remanufacturability Assessment (DRRA) tool is built based on the following four steps as illustrated in Figure 3.10.

Step 1: Select the features or aspects to be improved to facilitate remanufacturing, follow the list of remanufacturing design considerations and generate the feasible design alternatives.

Step 2: Adopt the Fuzzy TOPSIS method to evaluate and compare the impact of the design alternatives on the remanufacturing process.

Step 3: Evaluate the life cycle performance of the design alternatives using the proposed CED and Total Cost Analysis (TCA) method.

Step 4: Synthesize the results from step 2 and step 3 and make design decisions accordingly.

A computation tool has been developed based on the Visual C# to allow for fast computation and ease of use of this design aid, to assist decision makers in the application of the proposed methodology and to simplify computation complexity.

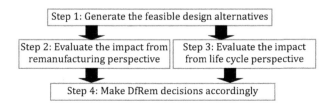

FIGURE 3.10 Flow chart for the proposed DfRem approach.

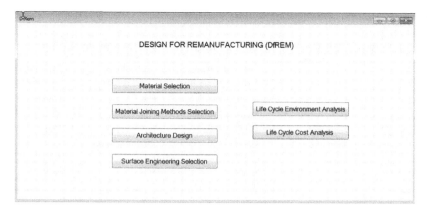

FIGURE 3.11   Screenshot of DfRem support tool.

With this tool, decision makers can obtain the impact of remanufacturing design features on both remanufacturing performance as well as the overall product life cycle, to improve the effectiveness and robustness of decision making and encourage greater incorporation of the remanufacturability concept during the product design stage. A screenshot of the program is shown in Figure 3.11. The screenshots for the sub-functions are shown in Appendix IV.

3.2.6 CASE STUDY I

In the following two sections, an engine block and an alternator have been selected to examine the applicability of the proposed design for remanufacturing support tool. The aim is to examine suitable design that can enhance the remanufacturability while improving the overall life cycle environmental performance (Yang et al., 2015a).

Engine block is the core of the engine and houses nearly all the components required for the engine to function properly. Many engine blocks in the early stage are manufactured from cast iron alloy due to its high strength and low cost. However, as engine designs become more complicated and heavier, some manufacturers have started to use lighter alloys, such as aluminum alloy, of which the density ratio to cast iron is 0.37 only. Due to relatively lower tensile strength and damping capacity of aluminum, such a design change may require a greater volume of aluminum to achieve a comparable functional performance of cast iron. Nevertheless, experience with practical substitution of cast iron with aluminum indicates that 1 kg of aluminum can replace up to 2 kg of cast iron for automotive product design (Vatsayan et al., 2014). Most recently, newly developed material processing technology has made compacted graphite iron (CGI) a viable alternative to gray cast iron for engine blocks. Therefore, in this case study, three different types of materials, namely, gray cast iron ASTM A48 Class 40, aluminum A356-t6 and CGI ASTM A482 Grade 450, are selected and their impacts on remanufacturing efficiency and life cycle performance are examined.

To evaluate the impact of the design alternatives on the remanufacturing efficiency, the methodology proposed in Section 3.2.3 will be adopted. The performance

TABLE 3.8
Candidate Materials for Engine Blocks and Their Performance Ratings

Criteria	Importance	Grey Cast Iron ASTMA48 Class 40	Aluminum A356-t6	CGI ASTM A482 Grade 450
Durability	High	Medium Good	Fair	Good
Cleanability	High	Medium Good	Fair	Medium Good
Restorability	Very high	Medium Good	Fair	Fair
EHS	Medium	Good	Medium good	Good
Cost	High	Good	Fair	Good
Complexity	N.A.	N.A.	N.A.	N.A.

TABLE 3.9
Relative Closeness and Ranking of Each Candidate Material

	Relative Closeness	Ranking
Grey Cast Iron ASTMA48 Class 40	0.82	2
Aluminum A356-t6	0.01	3
CGI ASTM A482 Grade 450	0.89	1

rating of the materials with respect to the six evaluation aspects and the subjective weights for each of the evaluation criteria are collected and shown in Table 3.8. Using the Fuzzy TOPSIS method, the ranking of the candidate materials in terms of their remanufacturability is calculated and shown in Table 3.9.

To evaluate the design alternatives from the life cycle perspective, the design information as well as the estimated functional and EOL performance of the three design alternatives is collected and presented in Table 3.10 (Adler et al., 2007; Vartabedian, 1992; Dawson and Indra, 2007; Smith and Keoleian, 2004). Detailed information, assumption and calculation can also be found in Yang et al. The parameters for energy intensity, such as the embodied energy intensity, manufacturing energy intensity and remanufacturing energy intensity, are stored in the design tool database, which would automate and speed up the calculation process (Koffler and Rohde-Brandenburger, 2010; Boustead and Hancock, 1979; Smith and Keoleian, 2004; Smil, 2008; Sutherland et al., 2008).

To demonstrate the environmental benefits of the design alternatives over complex life cycles, the energy consumption of aluminum and CGI engine block relative to cast iron engine block will be evaluated. The result of the environmental performance for the design alternatives using Equation 3.16 is shown in Figure 3.12, and the breakdown of energy consumption is shown in Figure 3.13.

TABLE 3.10

Specifications, Functional and EOL Performance of Design Alternatives

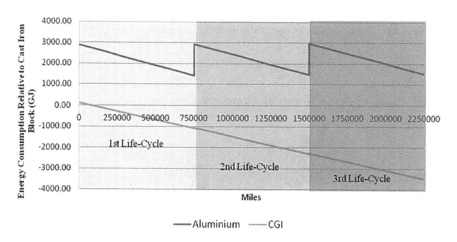

	Cast Iron Engine Block	Aluminum Engine Block	CGI Engine Block
Material	Grey Cast Iron ASTMA48	Aluminum A356-t6	CGI ASTM A482 Grade 450
Mass (kg)	158	130	134
Life mileage (km)	1,200,000	1,200,000	1,200,000
Functional performance	Equivalent	Equivalent	Equivalent
Reman rate (%)	60	50	70

FIGURE 3.12 Engine blocks life cycle environmental performance.

3.2.7 Case Study II

In this case study, the design evaluation was carried out for three alternative alternators designs. The three different alternator designs are shown in Table 3.11 (Schau et al., 2012). To evaluate their impact on the remanufacturing process, these three design alternatives are assessed using the evaluation criteria proposed in this framework. The performance ratings of the three alternators are shown in Table 3.12, and the calculated remanufacturing performance rankings are presented in Table 3.13.

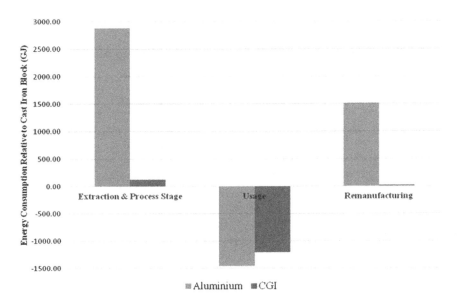

FIGURE 3.13   Engine blocks energy consumption breakdown.

TABLE 3.11
Specifications of Three Different Alternator Designs

	Design #I		Design #II		Design #III	
Component	Material	Mass (kg)	Material	Mass (kg)	Material	Mass (kg)
Belt fitting	Steel	0.52	Steel	0.52	Aluminum	0.18
Fan	Steel	0.14	Plastic/PP	0.02	Plastic/PP	0.02
Bearings	Rolled steel	0.10	Rolled steel	0.10	Plastic/PP	0.01
Housing	Cast Iron	2.53	Aluminum	0.96	Aluminum	0.96

To further evaluate their life cycle environmental performance, the design information and the estimated functional and EOL performance of the three design alternatives are shown in Tables 3.11 and 3.14, and they will be used as the input for the proposed design support tool. In this case, Design #I will be used as the reference to calculate the relative energy consumption of the other two design alternatives. The results of the study are shown in Figures 3.14 and 3.15.

TABLE 3.12
Design Candidates and Their Performance Ratings

Criteria	Importance	Design #I	Design #II	Design #III
Durability	High	Good	Medium good	Fair
Cleanability	High	Good	Medium good	Fair
Restorability	Very high	Good	Medium good	Fair
EHS	Medium	Good	Medium good	Fair
Cost	Low	Fair	Medium good	Good
Complexity	Low	Medium good	Fair	Fair

TABLE 3.13
Closeness and Ranking of Each Design Candidates

	Relative Closeness	Ranking
Design #I	0.91	1
Design #II	0.49	2
Design #III	0.09	3

TABLE 3.14
Functional and EOL Performance of Design Alternatives

	Design #I	Design #II	Design #III
Life mileage (km)	200,000	200,000	200,000
Functional performance	Equivalent	Equivalent	Equivalent
Belt fitting reman rate (%)	90	90	25
Fan reman rate (%)	90	0	0
Bearings reman rate (%)	50	50	0
Housing reman rate (%)	85	60	60

3.2.8 RESULTS AND DISCUSSION

In the first case study, CGI and GCI are ranked as the better material choices from the perspective of remanufacturing due to their desirable performance in wear resistance, fatigue resistance and reliability, which are the key enablers for successful engine block remanufacturing. Although aluminum demonstrates superior performance in "Density", it is a criterion considered less critical for product remanufacturability, thus leading to relatively lower rankings for aluminum. As from the life cycle environmental analysis, CGI engine block appears superior to aluminum and cast iron engine blocks due to its relatively lightweight design and satisfactory remanufacturing rate. Although the aluminum block design is the lightest among all the alternatives, it is comparatively more energy intensive to produce and remanufacture

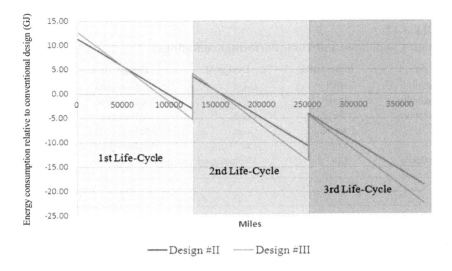

FIGURE 3.14   Alternators' environmental performance.

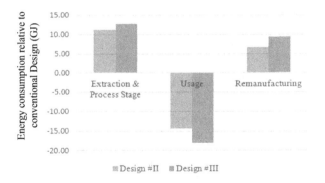

FIGURE 3.15   Alternators' energy consumption breakdown.

as observed in Figure 3.13, which overwhelms the benefits obtained from weight reduction energy savings.

In the second case study, Design #I is ranked as the best candidate from the remanufacturing perspective, due to its desirable performance on durability, cleanability and restorability, and hence it is much easier to be remanufactured into "good as new" condition. In comparison, as the plastic-based components are more fragile and prone to wear, this has made Design #II and Design #III less advantageous for remanufacturing. However, if the life cycle perspective is considered, Design #III has demonstrated its environmental benefit due to less energy consumed during the usage stage throughout the three life cycles, as shown in Figure 3.15.

The results of these case studies have highlighted the issue of using light duty materials in automotive design. Original equipment manufacturers tend to use lighter duty materials, such as aluminum or plastics, to reduce weight and subsequently improve the product performance (e.g., fuel efficiency and CO_2 emissions) during the use stage. However, this tends to make parts more fragile and/or prone to breakage during the remanufacturing processes, and thus reduces the number of remanufacturing cycles of the parts. In addition, failure due to the use of light duty material during the use stage often results in catastrophic destruction of components, which reduces the possibility of remanufacturing substantially. This issue has been addressed previously in a survey conducted in the automotive industry (Hammond et al., 1998) and is now validated quantitatively in this research. Moreover, it is known that many lightweight materials, e.g., aluminum, magnesium or polymer composite, are considerably more energy intensive to produce than, for example, conventional cast iron, prior to the use stage (Koffler and Rohde-Brandenburger, 2010). Hence, the benefits of the lightweight design strategy is dependent on whether the weight-induced energy saving during the use phase is sufficient to compensate for the potentially increased environmental impact of producing this part at the production stage as well as the remanufacturing stage. Therefore, a model which can estimate the overall environmental trade-offs of the different design concepts accurately within complex life cycles is of significance. The proposed tool has provided an easy-to-use and effective approach to support this analysis.

The results obtained for these two case studies are dependent on the underlying assumptions and data, among which, the successful remanufacturing rate, weight-induced fuel saving rate, life mileage and product life span are of major importance. To examine the impact of these situational variables on the results obtained, a sensitivity analysis was carried out. For example, sensitivity analysis indicates that the aluminum engine block would need a successful remanufacturing rate of 90% for two life cycles in order to achieve a better environmental performance than the cast iron engine block. Decision makers can adjust the values of the situational variables based on different constraints they faced and make decisions on the product design accordingly. Moreover, it is noted that the present studies have assumed the equality of the functional performance of the design alternatives in the use phase; future studies can account for the energy saving that might be induced by the changing of functional performance of design alternatives.

Recognizing that the design stage has possibly committed the greatest impact on EOL, the research topic of design for remanufacturing has received relatively generous amount of attention over the recent years. However, in reality, the increase in DfRem activity has yet to be realized proportionally. Therefore, factors that may affect the integration and implementation of DfRem, like management support, cross-functional communication, market demand, remanufacturing-related education and training, should be investigated properly (Hatcher et al., 2011). This part of the work can be referred to authors' work on "Towards implementation of DfRem into the product development process".

3.2.9 SUMMARY

To facilitate DfRem implementation, a holistic decision support tool is proposed in this section to steer a product design toward higher remanufacturability from four major design aspects, namely, material selection, material joining methods, structure design and surface coating. The impact of remanufacturability enhancement design features on both remanufacturing performance and overall product life cycle performance can be examined using the proposed MCDM methodology and CED/CC analysis, respectively, to achieve a robust and comprehensive remanufacturing design improvement.

The main contribution of this method is the compilation of the design features that are relevant to remanufacturing performance and the formulation of a systematic design tool for evaluating the design alternatives in a comprehensive and holistic manner. The tool can be adopted in the early design stage as only the relative remanufacturing performance ranking is required as the major input for remanufacturing impact analysis. Meanwhile, life cycle thinking is incorporated in the evaluation scheme, which further improves the effectiveness and robustness of the DfRem decision tool. The methodology has demonstrated quantitatively the issue of using light duty materials in automotive design for remanufacturing and highlighted the importance of design decision making from a life cycle perspective, which should include not only the manufacturing and use stage but also the EOL disposition.

3.3 SUMMARY OF THE CHAPTER

This chapter has discussed the approach to assess the feasibility of a product and its components for remanufacturing, and presented a holistic design support tool to improve the potential of the product for remanufacturing during product development stage. As mentioned previously, nearly 80% of the product cost is committed by the end of the product design stage. It is critical that the proposed remanufacturing activities in this chapter are carried out carefully and timely, if OEMs plan to incorporate remanufacturing as a part of their product life cycle. Through the proposed methodologies, the preliminary ideas of the EOL strategy of the product/components can be obtained, which will serve as necessary information to facilitate product remanufacturing decision making. To fit in with the already-sophisticated design process, the proposed design tool analyzes the "over-daunting" designs for remanufacturing guidelines and reorganizes them into four major design aspects, namely, material selection, material joining methods, structure design and surface coating, that designers are familiar with so as to reduce the burden of design for remanufacturing. Life cycle thinking has been incorporated into the design support tool that aims to deliver a robust and comprehensive design decision. These features have enabled proposed methodologies to be used effectively during the early design stage.

4 End-of-Life Stage

4.1 INTRODUCTION

With the increasing emphasis on environmental issues recently, treatment of EOL products is gaining attention. Even though a number of research areas on EOL determination and remanufacturability assessment have been identified, there are still some limitations from the following aspects. First, most of the research works assume a single quality grade for core return, ignoring the fact that the quality of the returned cores for remanufacturing is much more uncertain and dynamic than conventional manufacturing (Krikke et al., 1998; Song et al., 2005; Anityasari and Kaebernick, 2008; Jin et al., 2011). Second, as mentioned earlier, product remanufacturing decision making involves various dimensions of consideration (Goodall et al., 2014; Ziout et al., 2014), and there is still a lack of a holistic approach that can facilitate decision making during the remanufacturing process and ensure the completeness of operational, technological, economic and environmental considerations. Further, a methodology that considers both complete and partial disassembly and optimizes the EOL decision for each component in a product stewardship system has yet to be fully delineated. These limitations and gaps have constituted to the contributions as well as the motivations of the following research.

In this chapter, a decision support framework for EOL decision making for components of a returned product type is proposed. The framework includes both qualitative and quantitative analyses to address the operational and technological considerations for product remanufacturing while optimizing the environmental and economic performance. Probability theory is utilized in this framework to analyze the impact of the quality of the returned products on EOL decision making. In addition, to represent a product structure hierarchy and the interconnections among the components of a product, the HALG is used, allowing both complete and partial disassembly strategies to be considered during EOL strategy planning.

4.2 RECOVERY DECISION MAKING FOR COMPONENTS OF RETURNED PRODUCTS

Recovery Strategies Definition

Common EOL strategies include reuse, remanufacturing, recycle, landfill and incineration. Considering that the proposed decision support system is used for companies whose main focus is on product remanufacturing, the following EOL strategies will be considered:

Upgrade: The component will be upgraded with the state-of-the-art technology to improve its performance or quality to be on par with the latest standards, so as to meet the market demand.

Restore: The component will be returned to "as good as new" conditions. The options of upgrade and restore are mutually exclusive.

DOI: 10.1201/9781003275596-4

Disposal: The component will be disposed of and replaced with a new component; material which has recovery value will be recycled and the rest of the material will be incinerated or landfilled.

4.2.1 HALG FOR PRODUCT STRUCTURE REPRESENTATION

The HALG (Dong et al., 2006) is used to represent the product structure hierarchy of a returned product; it can represent the interconnections of the components and subassemblies, which leads to the ease of considering partial disassembly during the disassembly process. In an HALG, the squares and discs represent the subassemblies and components, respectively, and the arcs represent the connections between components and subassemblies. Figure 4.1 shows the HALG representation for an automotive alternator. Hence, the alternator comprises four levels, four subassemblies and nine components. For each subassembly, there can be different disassembly strategies, which can produce two types of components, namely, *Independent component*, where all the joints connected to that component are disconnected, and the *Connected component*, where there are remaining joints on that component (Yang et al., 2015b).

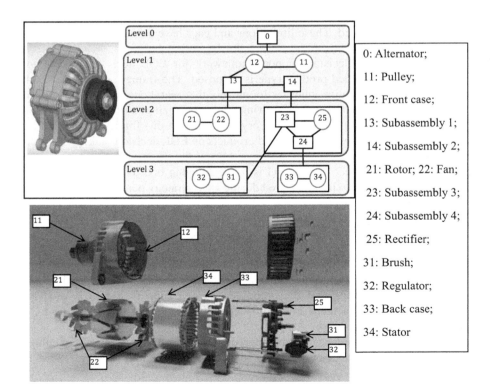

FIGURE 4.1 HALG representation for an alternator (Kim et al., 2008; Schau et al., 2012).

4.2.2   Operational and Technological Assessments

The objectives of the operational and technological assessments are to remove the subassemblies or components that are obviously non-reusable and classify the reusable subassemblies and components into different quality levels. These assessments consist of the following three steps.

Step 1: A visual and physical inspection is conducted to identify severely damaged and worn items and discard them, e.g., subassemblies that are severely worn, components that are usually replaced, etc. This process can vary among the types of cores being inspected and is usually performed manually without any tools or instruments.

Step 2: This step estimates the demand of the components and subassemblies. Components and subassemblies that have no demand will be disposed of at this stage. The estimated demand and information of the components and subassemblies allow inventory to be managed for future use.

Step 3: This step assesses the functional performance and assigns a quality level to a subassembly or component. The returned cores usually have varying quality levels. The product quality standards are usually set by the OEM, international standards or remanufacturing firms. Based on these standards, the quality of the components will be examined and graded. Besides identifying the quality level for each subassembly or component, a subassembly or component that has been tested to be non-reusable will be discarded.

In this framework, the focus is not on the specific mechanisms for quality inspection in these steps as they often vary among the remanufacturers. It is the assessment that must be carried out and the quality levels identified from the technical assessments which are of interest to this methodology.

4.2.3   Conditional Probability of Quality and Expected Profit

The quality of the returned products influences the recovery strategy, and it has been identified that remanufacturing cost decreases as the quality increases (Ferguson et al., 2009). In this research, for the purpose of simplicity, two quality levels will be used, namely, "good conditions" ($q=1$) or "malfunctioning" ($q=2$).

During a disassembly process, there is uncertainty in the quality of the released components and subassemblies. This uncertainty is modeled and defined as a conditional probability $\text{Pr}_{ij}\,(q1|q2)$, which represents the probability of the quality of its subcomponent i equals to $q1$, given the quality level $q2$ of assembly j.

Assuming $C_{i+1,j}$ represents the component released from subassembly SA_{ij} and $\text{PF}\left(C_{i+1,j}, q_{i+1,j}\right)$ represents the profit of each component $C_{i+1,j}$ under quality level $q_{i+1,j}$, by using the concept of conditional probability of the quality level, the expected profit $\text{PF}\left(\text{SA}_{ij}, q_{ij}\right)$ for processing the subassembly SA_{ij} with quality level $q_{i,j}$ is calculated using Equation 4.1, where DC_{ij} represents the disassembly cost associated with subassembly SA_{ij}.

$$\text{PF}\left(\text{SA}_{ij}, q_{ij}\right) = \sum_{C_{i+1,j}} \sum_{q_{i+1,j}} \text{Pr}_{C_{i+1,j}}\left(q_{i+1,j} \mid q_{ij}\right) * \text{PF}\left(C_{i+1,j}, q_{i+1,j}\right) - \text{DC}_{ij} \qquad (4.1)$$

4.2.4 Economic and Environmental Indices

Several objectives can be considered for planning the optimum EOL strategy of a component. In this research, the objective is to maximize the economic performance while minimizing the environmental impact. An economic index and an environmental index will be calculated, where all the variables will be evaluated on a monetary scale, and equal weighting will be assumed for each objective.

The economic effectiveness of remanufacturing is determined by benchmarking with EOL options of disposal and replacing with new components in the context of the OEM remanufacturers. Equation 4.2 is used to calculate the component economic index. The cost of component upgrading or restoring is calculated from the remanufacturing processing cost, and the cost of the disposal option is calculated from the component's disposal cost plus the cost of producing or ordering the replaced components. The consideration of replacement cost is of great necessity in a product stewardship system, as part of the replacement cost may either encourage or discourage the remanufacturing decision. For example, if the replacement cost is prohibitively high, component remanufacturing is usually recommended; otherwise, replacing the low-cost components with new ones might be more economical.

The environmental impact of remanufacturing is assessed based on material consumption, energy consumption, waste generation and toxicity discharged, which are factors/categories used commonly for analyzing the environmental performance (Smith and Keoleian, 2004). These measurements will be converted into monetary values, so as to be comparable with economic performance calculation. Even though there are other environmental impacts, such as the loss of non-renewable resources, greenhouse effects, the impacts of these factors are currently not required by laws to be borne by companies, and therefore they are excluded in this study. However, the proposed method can be extended by adding any environmental impact factor/category that is required in certain applications. In addition, the environmental impact of the disassembly operations is neglected since most of the disassembly operations are performed manually. Equation 4.3 shows the calculation of the environmental index. Thus, the overall index of component ij is calculated using Equation 4.4, which is a summation of the economic and environmental indices.

Economic index:

$$\left(EC_{ij}, X_{ij} \right) = \left(PC_{ij}, X_{ij} \right) + \left(NC_{ij}, X_{ij} \right) \tag{4.2}$$

Environmental index:

$$\left(EV_{ij}, X_{ij} \right) = \left(RMR_{ij}, X_{ij} \right) * RMC_{ij} + \left(EP_{ij}, X_{ij} \right) * EC + \left(WDP_{ij}, X_{ij} \right) * WMC_{ij}$$

$$+ \left(TOX_{ij}, X_{ij} \right) * TMC_{ij} \tag{4.3}$$

Overall index for component ij:

$$PF\left(C_{ij}, X_{ij} \right) = \left(EC_{ij}, X_{ij} \right) + \left(EV_{ij}, X_{ij} \right) \tag{4.4}$$

where

ij: The jth component on the ith level;

X_{ij}: Indicator of the EOL strategy of component ij;

$X_{ij}=1$, if the component ij is to be upgraded;

$X_{ij}=2$, if the component ij is to be restored;

$X_{ij}=3$, if the component ij is to be discarded with replacement;

$\left(EC_{ij}, X_{ij}\right)$: The economic index of component ij, if EOL option X_{ij} is taken;

$\left(PC_{ij}, X_{ij}\right)$: The cost of restoring or upgrading the component ij (if $X_{ij}=1$ or $X_{ij}=2$) or the cost of disposing the component ij (if $X_{ij}=3$) (\$);

$\left(NC_{ij}, X_{ij}\right)$: The cost of producing or ordering the new component ij when $X_{ij}=3$; otherwise $\left(NC_{ij}, X_{ij}\right)=0$ (\$);

$\left(EV_{ij}, X_{ij}\right)$: The environmental index of component ij, if EOL option X_{ij} is taken;

$\left(RMR_{ij}, X_{ij}\right)$: Mass of raw material required for restoring/upgrading the component ij (if $X_{ij}=1$ or $X_{ij}=2$) or replacing with a new component ij (if $X_{ij}=3$) (kg);

RMC_{ij}: Raw material cost (\$/kg);

$\left(EP_{ij}, X_{ij}\right)$: Energy required for restoring or upgrading the component ij (if $X_{ij}=1$ or $X_{ij}=2$) or disposal and replacing with a new component ij (if $X_{ij}=3$) (MJ);

EC: Energy cost (\$/MJ);

$\left(WDP_{ij}, X_{ij}\right)$: Waste generated for restoring or upgrading component ij (if $X_{ij}=1$ or $X_{ij}=2$) or disposing the component ij and replacing with a new component (if $X_{ij}=3$) (kg);

WMC_{ij}: Waste management cost (\$/kg);

$\left(TOX_{ij}, X_{ij}\right)$: Toxic discharged during restoring or upgrading of component ij (if $X_{ij}=1$ or $X_{ij}=2$) or disposing the component ij and replacing with a new component (if $X_{ij}=3$) (kg);

WMC_{ij}: Toxicity management cost (\$/kg);

$PF\left(C_{ij}, X_{ij}\right)$: The overall index of component ij, if EOL option X_{ij} is taken.

4.2.5 OVERALL APPROACH FOR EOL DECISION MAKING OF RETURNED PRODUCTS

An optimum combination of the EOL strategies for the components of a returned product is determined based on the following three steps that are illustrated in Figure 4.2.

Step 1: Conduct operational and technological assessments (Section 4.2.3: Steps 1 and 2) to identify the subassemblies or components that are non-reusable. Develop the HALG (Section 4.2.2) for the remaining subassemblies and components.

Step 2: Stochastic dynamic programming to determine an optimum EOL strategy for each component by

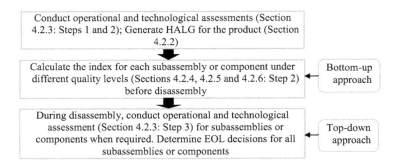

FIGURE 4.2 Flow chart for the EOL strategy planning.

a. Estimating the overall index for each component for different EOL strategies under different quality levels, according to Equations 4.2–4.4 (Section 4.2.5).
b. Starting from the lowest level i (L_i), for each component ij (C_{ij}) and for each quality level (Q_{ij}), choose an EOL strategy $X_o \in X_{ij}$, such that the overall index of component ij is maximum $\text{MPF}(C_{ij}, q_{ij})$, i.e.:

$$\text{MPF}(C_{ij}, q_{ij}) = \max_{X \in X_{ij}} \text{PF}(C_{ij}, q_{ij}, X) \tag{4.5}$$

$$\text{Set } X^{\text{best}}(C_{ij}, q_{ij}) = X_0 \tag{4.6}$$

c. If $i > 1$, $i = i - 1$
   c1: For each component (C_{ij}) in level i (L_i):
   For each quality level (Q_{ij}), choose an EOL strategy $X_o \in X_{ij}$, such that the overall index of component ij is maximum ($\text{MPF}(C_{ij}, q_{ij})$), i.e.:

$$\forall C_{ij} \in L_i, \forall q_{ij} \in Q_{ij}; \text{MPF}(C_{ij}, q_{ij}) = \max_{X \in X_{ij}} \text{PF}(C_{ij}, q_{ij}, X) \tag{4.7}$$

$$\text{Set } X^{\text{best}}(C_{ij}, q_{ij}) = X_0 \tag{4.8}$$

c2: For each subassembly (SA_{ij}) in level i (L_i):
Enumerate all the disassembly strategies D_{ij} for SA_{ij}. For each disassembly strategy $d \in D_{ij}$, label the subcomponents ($C_{i+1,j}$) released from SA_{ij} as follows:

- **Independent component** ($\text{ID}\_C_{i+1,j}$), if all the joints connected to component $C_{i+1,j}$ are disconnected. The maximum overall index of each independent component is equal to the maximum overall index $\text{MPF}(C_{ij}, q_{ij})$ calculated from Step b or c1.
- **Connected component** ($\text{CN}\_C_{i+1,j}$), if there are remaining joints on component $C_{i+1,j}$. The maximum overall index of this group of components is equal to the sum of the indices of disposing these components, i.e., $\sum_{\text{CN}\_C_{i+1,j}} \text{PF}(\text{CN}\_C_{i+1,j}, X = 3)$ as disposal is the only feasible EOL strategy for these connected components.

Therefore, for each quality level (Q_{ij}), and for each disassembly strategy D_{ij}, choose the disassembly strategy $d_o \in D_{ij}$ such that the overall index of the subassembly SA_{ij} is maximum, i.e.,

$$\forall SA_{ij} \in L_i, \forall q_{ij} \in Q(ij), \forall d \in D_{ij}$$

$$\mathrm{MPF}\left(SA_{ij}, q_{ij}\right) = \max_{d \in D_{ij}} \left\{ \sum_{\mathrm{ID}\_C_{i+1,j}} \sum_{q_{\mathrm{ID}\_C_{i+1,j}}} \Pr_{\mathrm{ID}\_C_{i+1,j}} \left(q_{\mathrm{ID}\_C_{i+1,j}} \middle| q_{ij}\right) * \mathrm{MPF}\left(\mathrm{ID}\_C_{i+1,j}, q_{\mathrm{ID}\_C_{i+1,j}}\right) \right.$$

$$\left. + \sum_{\mathrm{CN}\_C_{i+1,j}} \mathrm{PF}\left(\mathrm{CN}\_C_{i+1,j}, X = 3\right) - \mathrm{DC}(d) \right\} \tag{4.9}$$

$$\text{Set } d^{\mathrm{best}}\left(SA_{ij}, q_{ij}\right) = d_0 \tag{4.10}$$

Noted that if $d_0 = [\varnothing]$, the best disassembly strategy is to dispose the subassembly SA_{ij}, as a whole without further disassembly.

d. If $i > 1$, $i = i - 1$, go to Step c, else go to Step e.

e. The stochastic dynamic programming stops at level 0 (product level).

Step 3: During the disassembly process, if the quality level needs to be examined, operational and technological assessments will be performed (Section 4.2.3: Step 3). Otherwise, the optimum EOL options determined in Step 2 will be adopted. The assessment will stop when an EOL option has been determined for each of the components or subassemblies.

4.3 CASE STUDIES

Two case studies have been chosen and conducted to demonstrate the proposed methodology, which are an automotive alternator and a hedge trimmer.

4.3.1 CASE STUDY I

Alternators, which are basic automotive parts, are chosen to illustrate the applicability of the proposed model. Remanufacturing of alternators is comprised of more than 45% of the revenue in the North American aftermarket and is performed by hundreds of companies in 2005. To apply the proposed methodology for an alternator, first, the operational and technological assessments are performed to remove components that are non-reusable or frequently replaced, such as washers, screws, springs. After that, the HALG for the remaining components is constructed and shown in Figure 4.1, which allows the users to visualize the hierarchy of the product structure and the interconnections among the components, and consequently allowing the partial disassembly strategies to be considered and interpreted during EOL strategy planning.

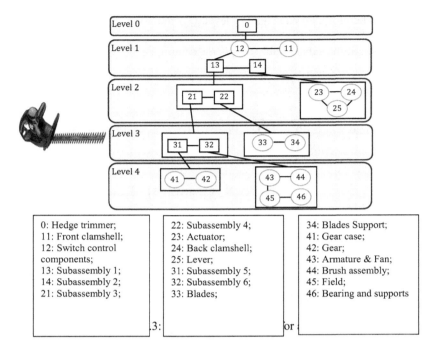

FIGURE 4.3   EOL strategy planning for a trimmer.

The required inputs for the model, e.g., economic and environmental index, disassembly cost and conditional probability are estimated and given in Appendices V–VIII (Kim et al., 2008; Schau et al., 2012). The result of applying the proposed methodology, namely, using Equations 4.5–4.10, to plan the EOL strategies of an alternator is represented graphically in Figure 4.3. Quality inspection is performed at points which could lead to different EOL decisions. Specifically, quality inspection at the subassembly level, e.g., SA14, would mean that different disassembly strategies might be suggested under different quality states, whereas quality inspection at the component level, e.g., C25, implies that if the quality of the component is good, remanufacturing is recommended, otherwise, disposal and replacing with new component is recommended.

4.3.2 CASE STUDY II

The proposed methodology is further utilized to examine the EOL options for a power tool, i.e., a hedge trimmer, which is remanufactured by some OEM companies, such as Robert Bosch (Atasu et al., 2010). In cases where the economics of remanufacturing these products may be marginal (or perhaps negative), there are still several strategic reasons to consider offering these remanufactured products, such as recovering costs from commercial returns, fending off competition from independent third-party competitors, commitment to corporate social responsibility.

To determine the product structure and component information, a hedge trimmer was disassembled completely. An HALG was constructed accordingly and shown in Figure 4.4. Appendices IX–XII summarize the input data used for this case study. Through applying the proposed methodology, the suggested EOL disposition for a hedge trimmer is graphically described in Figure 4.5.

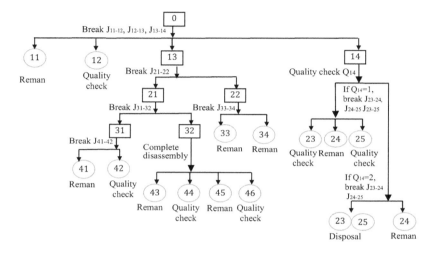

FIGURE 4.4   HALG representation for a hedge trimmer.

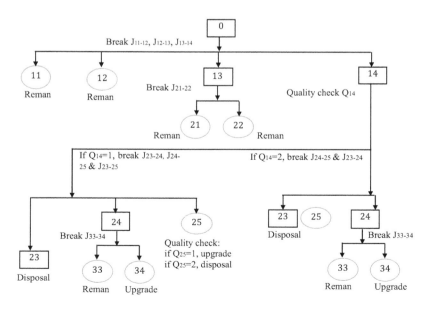

FIGURE 4.5   EOL strategy planning for a hedge trimmer.

4.4 RESULTS AND DISCUSSION

The proposed methodology comprises three phases, which involve both qualitative and quantitative analyses to ensure the completeness of operational, technological, economic and environmental decisions. In the first phase, the simplest and quickest inspection is conducted so that many obviously non-usable subassemblies or components are rejected and to ensure only useful items proceed to the next decision-making stage. This phase is meant to establish if it is possible and/or necessary to remanufacture a given core or component, from operational and technological perspectives. For example, washers, screws and springs of the alternator and the hedge trimmer are identified in this phase as "non-usable" components, and therefore are excluded from further examination. In the second phase, a quantitative analysis from the economic and environmental aspects is conducted through dynamic programming in a bottom-up manner. The results from this analysis are the quality-dependent EOL strategy as well as the suggested disassembly depth as shown in Figures 4.3 and 4.5, which constitute the basis and knowledge to assist the actual remanufacturing decisions. The final EOL decision can only be made during the actual disassembly process, where the quality of the subassembly is inspected, and sufficient information has been gathered. In the third step, quality inspection is performed at points which would lead to different EOL decisions being made, e.g., quality check is required for 14 and 25 of the alternators. It should be noted that during the reconditioning process, "remanufacturable components" will be continuously inspected until they have been determined to be accepted or have failed, to ensure the quality of the remanufactured product for the next life cycle, which, however, is not the focus of this study. The results of the two case studies have showcased the applicability of the proposed methodology for determining the disassembly and recovery strategy that should be carried out to handle the flow of returned products.

As mentioned previously, the quality of the returned product would influence the recovery strategy. Due to the lack of information before disassembly, the quality state of the components released in a disassembly step usually involves uncertainty. To deal with this uncertainty and maximize the economic and environmental rewards of every disassembly step, a quality classification scheme, transition probability and expected value calculations are employed. The results are a complete set of conditional assignment rules as shown in Figures 4.4 and 4.5. For example, if subassembly 14 of the hedge trimmer is in "good condition" ($q = 1$), a complete disassembly strategy will be suggested, otherwise, only component 24 will be extracted from the subassembly, leaving the rest of the components to be disposed as a whole without further disassembly. Note that in this study, the quality of a return flow is reflected using simplified technical states, namely, "good condition" ($q = 1$) or "malfunctioning" ($q = 2$). However, the classification could be further defined with more quality grades or the aspect of classification could be expanded to include composition, usage condition and quantity to account for greater uncertainty or possibilities that might be involved in EOL decision making.

The results of the case studies have also demonstrated the importance of considering disassembly cost during EOL strategy planning. For example, in the alternator case, if the regulator (with $q = 1$) is considered independently, remanufacturing might be a viable strategy than disposing and replacing. However, if the disassembly cost between the brush and the regulator is considered, it would be a better strategy to dispose and replace the regulator and the brush together without further disassembly. The insight gained from this result can be used as design feedback to facilitate product design for remanufacturing, so as to reduce the disassembly cost and make the product more viable for remanufacturing.

In the proposed methodology, the optimization objective is to maximize the economic surplus and minimize the environmental impact. To illustrate the impact of the objectives on the results obtained, two different objectives are investigated in this section. It is observed that when the objective is solely to maximize the economic aspect, some recovery decisions will become quality dependent, e.g., subassembly 21, 22, 31, 32 of the hedge trimmer. If the quality of the subassembly is good, disassembly is recommended, otherwise, direct disposal and replacing with new component is recommended. Figures 4.6 and 4.7 illustrate these results. When the objective is to minimize the environmental impact, a complete disassembly strategy is recommended to retain the value of the components regardless of the cost involved. Figures 4.8 and 4.9 illustrate these results. Therefore, the EOL decisions depend on the objective(s) set by the users, e.g., the company management, and the proposed methodology has provided an effective and quantitative approach for comparing environmental impact with economic consideration.

A computation tool based on the MS Excel@ platform has been developed to implement the proposed methodology. Hence, instead of relying on *ad hoc* engineering judgment, decision makers can refer to the proposed tool for the optimum EOL decision, which corresponds to maximum economic and environmental rewards and the necessary depth of disassembly to avoid a waste of time and effort on separating the components that could have been disposed together. Meanwhile, the addition

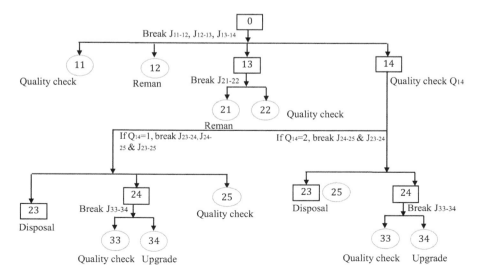

FIGURE 4.6   Results considering economic impact only (alternator).

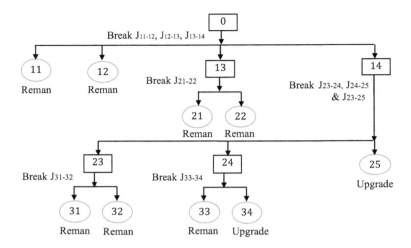

FIGURE 4.7   Results considering economic impact only (hedge trimmer).

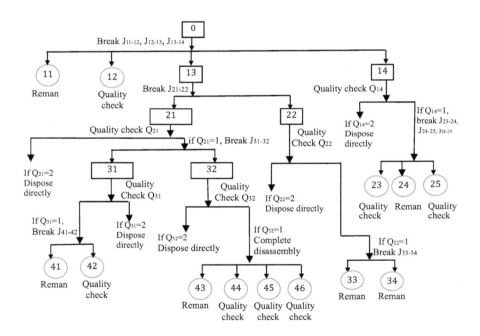

FIGURE 4.8   Results considering environment impact only (alternator).

of the qualitative analysis, which incorporates the technological and operational considerations, has further ensured the comprehensiveness and reliability of decision making. These characteristics make the proposed tool potentially applicable for determining the tactical EOL strategies for different types of returned products.

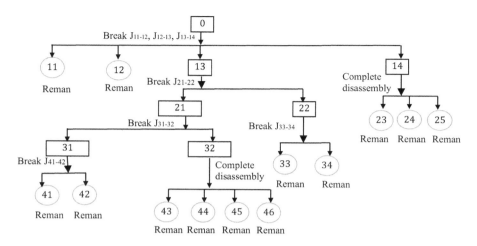

FIGURE 4.9 Results considering environment impact only (hedge trimmer).

Note that the proposed methodology assumes that products are dealt with in a decoupled way and the financial synergy resulting from sharing set-up costs among multiple products is not considered. However, if the results of one-product optimization are used on a multiple-product level, it is necessary to iterate the proposed methodology with the readjusted costs and revenues to recalculate the EOL strategy on the single-product level. These issues are subject to further research.

4.5 SUMMARY OF THE CHAPTER

A methodology is proposed in this chapter to facilitate the EOL decision making for components of returned products. This model makes contributions in: (1) considering the quality of the components and returned product by using the concept of conditional probability; (2) using both qualitative and quantitative analysis to ensure the completeness of operational, technological, economic and environmental decisions; and (3) adopting HALG to represent the product structure and the interconnections among components, and thus determine the optimal depth of disassembly and the EOL option for each component. The proposed model is universal in application and can be adapted for different product types by adjusting the data and variables involved in the model. For further research, disassembly sequencing planning can be integrated with the proposed methodology to further improve the disassembly efficiency and reduce the disassembly cost. The technical status of the components can be defined in more dimensions, e.g., composition, usage condition, quantity, etc., to provide more information for EOL strategy planning. Artificial intelligence techniques, e.g., fuzzy logic, can be applied to address the subjectivity of the decision-making process and improve the reliability of the proposed methodology.

5 Integrative Solution Guide for Remanufacturing and Opportunities for Industry 4.0 to Support Remanufacturing

5.1 DISCUSSION OF THE DRRA FRAMEWORK AND CASE ILLUSTRATION

The primary objective of this solution guide is to develop a Design for Remanufacturing and Remanufacturability Assessment (DRRA) tool that is to be used at the early product development stage to assess the suitability of a product and its components for remanufacturing and make suitable modification of the design to improve the product potential for remanufacturing. The DRRA tool can be used during the product return/service stage to assist in the generation of the recovery plan of the returned products taking into account their quality variation. The methodologies are connected internally by feeding forward and backward the product knowledge, such as product design features, EOL strategy planning and remanufacturing considerations. The flow of the product knowledge, along with the proposed decision support tools, has formed collectively the proposed DRRA tool, as illustrated in Figure 1.2. The flow of the product knowledge between the proposed methodologies is discussed and illustrated with an alternator case study.

5.1.1 FEED FORWARD OF THE PRODUCT KNOWLEDGE FROM THE EARLY-STAGE EOL STRATEGY PLANNING TO EOL STAGE RECOVERY PLAN GENERATION

Using the EOL strategy planning tool presented in Section 3.1, all the necessary factors which affect remanufacturing will be considered and the subassemblies and parts that are feasible candidates for remanufacturing can be identified successfully. Using an alternator as an example, components such as stator, rotor, housing and fan can be easily identified as reusable components for remanufacturing using the remanufacturability screening test. On the other hand, due to limited remaining useful life and remanufacturing capability, components such as slip rings, springs, washers and screws are classified as non-feasible candidates for remanufacturing and excluded for further remanufacturing analysis. This EOL information, together with the bill of materials, disassembly instruction and repair manual, will form the

DOI: 10.1201/9781003275596-5

valuable product knowledge and be fed forward to the EOL stage. Remanufacturers, during the product EOL stage, can employ this information to conduct operational and technological assessments to extract the reusable components for remanufacturing, develop the Hierarchical Attributed Liaison Graph (HALG) for the alternators and generate a preliminary recovery plan for the returned type of the alternator, as illustrated in Sections 4.3 and 4.4. This, to a certain extent, can reduce the effort of remanufacturers to re-establish the product knowledge which already exists.

5.1.2 FEED FORWARD AND BACKWARD OF THE PRODUCT KNOWLEDGE BETWEEN EARLY-STAGE EOL STRATEGY PLANNING AND EARLY-STAGE DESIGN FOR REMANUFACTURING

Besides assessing the remanufacturability of products and components at the early design stage, this research takes a proactive measure to improve the potential of products and components for remanufacturing through the development of an effective and efficient product design support tool. As this design tool is meant to be applied to the components or parts that have the potential for remanufacturing, it would require extracting EOL-related information from the product EOL strategy planning tool. Using the alternator as an example, when its reusable components, such as housing, belt fitting, fan and bearing, have been identified using the proposed EOL strategy planning tool, designers can make use of this information to proceed with design for remanufacturing. As illustrated in case study 2 in Section 3.2, designers can vary the materials selection for the selected components and examine their impact on both remanufacturing efficiency and life cycle performance. On the other hand, whenever there is any design modification, the redesign specification can be sent back to the product EOL strategy planning tool to go through remanufacturability analysis or re-run with the optimization model to examine their impact on EOL recovery strategy. Details of this can be seen in Section 3.1.

5.1.3 FEED BACKWARD OF THE PRODUCT KNOWLEDGE FROM EOL REMANUFACTURING STAGE TO EARLY-STAGE PRODUCT DESIGN FOR REMANUFACTURING

During the remanufacturing process, remanufacturers will establish their own product knowledge, which in many cases complements and even overcomes the available product information. For example, in the alternator case study in Section 4.3, if the joining method between the regulator and the brush can be designed to facilitate the disassembly process, it would be more viable to remanufacture the regulator rather than dispose and replace it. This kind of experience of the remanufactured product and operations, along with documentation of the core quality and defect measurement, composes the valuable product knowledge, which can be fed back to the early design stage to facilitate product design for remanufacturing. When designers receive this information, they will have a better understanding about the design features that might impede or facilitate the remanufacturing process and address the remanufacturing concern through proper product feature design.

The increase in product knowledge, through the feeding forward and backward of the product information, contributes toward a more efficient DRRA system, which consequently would promote transparent and accessible product life cycle information flow and stimulate the product remanufacturing development.

5.2 REMANUFACTURING RIDES THE WAVE OF INDUSTRY 4.0

Industry 4.0 refers to the fourth stage of industrialization that aims for a high level of automation in the manufacturing industry through adoption of ubiquitous Information and Communication Technology. The advent of Industry 4.0 has presented immense opportunities to unlock the potential of remanufacturing. In this section, the opportunities brought by Industry 4.0 for remanufacturing will be discussed based on the three aspects, namely, smart life cycle data for design for remanufacturing and EOL management, smart factory for cost-effective and green remanufacturing operations, and smart services for a successful remanufacturing business model. Figure 5.1 shows the three application areas and the technology enablers from Industry 4.0.

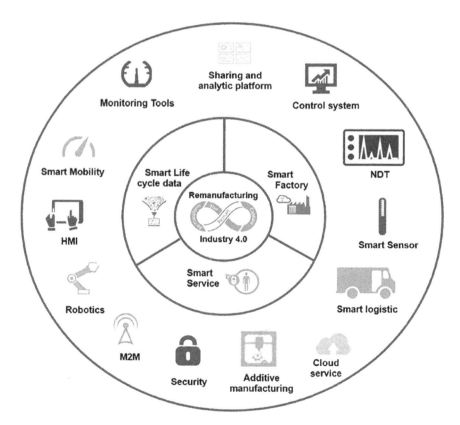

FIGURE 5.1 Opportunities from Industry 4.0 for Remanufacturing and its key enablers (Yang et al., 2018).

The technology enablers are presented in the outer rim of the circle and include smart sensors, cloud computing, robotics, machine to machine communication and additive manufacturing.

5.2.1 SMART LIFE CYCLE DATA

From the initial product design and development stage to the EOL stage, various types of product information are generated and captured. Ideally, this information should be shared across the stakeholders to support product life cycle management. However, the flow of product information remains essentially unestablished due to ineffective data extraction, loss of information during product transfer between stakeholders, undeveloped platforms to support information sharing and other policy restrictions. Ineffectiveness in data circulation has reduced the efficiency of product life cycle management and the quality of service provided. For example, incomplete information on returned cores remains a big challenge for the majority of remanufacturers. To restore these products back to "good as new" quality, remanufacturers have to recreate product knowledge which already existed at the product design stage. In this regard, the digital transformation in Industry 4.0 has shed some lights on addressing this concern through improving data transferability and building the knowledge/data sharing platform. This could be enabled through sensors, embedded systems, connected devices ("Internet of Things") as well as a comprehensive data management platform. For example, when information regarding computer-aided design, bill of materials, parts information, manufacturing and assembly instructions, data from the product use stage and historical repair information are stored in a central system and easily accessible by remanufacturers, such that the repair decisions during the remanufacturing process could be made easily and operations could be carried out in a more efficient manner. Furthermore, when information, such as product failure modes and rates, replacement frequency, cleaning efficiency, disassembly challenges and upgrading challenges, is extracted effectively from the remanufacturing stage and fed back to product designers, many of the barriers occurring during the remanufacturing process could be avoided for the next generation of products through incorporating proper design features. This would be strategically important and a substantial cost-saving measure, as more than 70% of product costs are determined at the product development stage.

5.2.2 SMART FACTORY

Due to the uncertainty involved in the quantity and quality of cores returned, remanufacturing operations need to incorporate a high degree of flexibility to react quickly and appropriately to the various product reconditioning requirements. Smart factories, which enable high flexibility and small batch size production, seamlessly address this complexity issue associated with remanufacturing operations. Smart factories are essentially at the core of Industry 4.0. Smartness is achieved using electronic hardware/software as well as networking of production resources. Compared with traditional manufacturing, more ancillary hardware and software, like radio-frequency identification (RFID) tags, barcodes, laser markers, sensors and communication infrastructure, will be embedded into a factory to enable machines to collaborate with each other using intelligent analytics. In future, in a smart remanufacturing environment, machines could obtain incoming core information through scanning a barcode attached to the core, adapt the remanufacturing operations through self-optimization and smart fixturing capabilities, update the process-related information to a database via wireless transfer, store remanufacturing knowledge gained from experience, etc. This could enable a substantial reduction of the labor force and lessen the dependency on highly skilled operators. Meanwhile, with various types of sensors embedded into equipment and cells, data from the manufacturing process could be retrieved and sent for real-time analysis. This would support the early detection of machines or cell failures and allow preventative strategies to be implemented to avoid unplanned maintenance and catastrophic failures. Information associated with the product manufacturing data can be recorded and stored as part of the product life cycle data. In addition, energy-efficient remanufacturing processes can be achieved through collecting real-time energy consumption data and implementing an energy management system accordingly.

Innovative technologies, such as additive manufacturing, 3D scanning, automated-guided vehicles, inspection drones and augmented reality technologies (Chang et al., 2020), will continuously drive down the cost of remanufacturing operation while delivering substantial improvements in the quality of the repaired products. Using aerospace remanufacturing as an illustration, worn airfoils can sometimes be repaired using the laser metal deposition technique, which is an additive process where metal powder is melted by a Computer Numerical Control (CNC) laser on a robot arm to form a direct fusion-bonded deposit on the blades. This technology offers the precision and low heat input necessary to successfully achieve such restoration, compared to more conventional fusion welding processes.

5.2.3 SMART SERVICES

One of the challenges that remanufacturers face is the control of the timing, quality and quantity of cores returned. In this regard, product service systems (PSSs) have provided opportunities to cope with the complexity of core return for remanufacturing. In this emerging and disruptive business model, the ownership of a product is usually retained by the Original Equipment Manufacturers (OEMs) or retailers and only the service or usage is offered to customers (e.g., selling "flying hours of the engine" instead of selling "engines"). Hence, it creates a mandate for manufacturers or retailers to monitor their product's performance during its runtime and to forecast remanufacturing operations on the core returned based on the predicted remaining useful life of the product. On the other hand, from the consumers' perspective, as they pay for the service rather than the ownership of the product, market acceptance for remanufactured goods will likely be increased, leading to a successful remanufacturing model. Real-time monitoring of product in-use and data analysis via embedded sensor networks and cloud-based computation could enable predictive maintenance with the early detection of problems. Increased connectivity among products, customers and manufacturers, promoted by Industry 4.0, presents immense opportunities for boosting the product service model. Using machines that are leased to the power plant sector as an example, power plants rely heavily on the continuous availability of their machinery; to increase the reliability of machines, smart sensors are embedded into the machines to monitor in real-time critical factors, such as temperature, pressure, switches, energy consumption and vibration. Sensor data are collected and logged to a central device through a network for prediction of potential wear, estimation of components' useful life and to schedule maintenance or remanufacturing of components, in a timely manner.

5.3   SOLUTIONS CONTRIBUTIONS

The contribution and novelties of the solution guide are summarized as follows:

a. Analyzing the remanufacturability of a product and its components at the early product development stage, by addressing the comprehensive aspects of remanufacturing considerations and utilizing NSGA-II to determine explicitly a Pareto set of optimal EOL strategies;

b. Investigating a large number of scenarios as necessary in EOL strategy planning, such as changes in remanufacturing cost, product design, landfill cost and suggest appropriate strategies to accommodate the changes in the situational variables or measures to improve product design;

c. Steering a product design toward higher remanufacturability from four major design aspects, namely, material selection, material joining methods, structure design and surface coating;

d. Examining the impact of remanufacturability enhancement design features on both remanufacturing performance and the overall product life cycle performance using Fuzzy TOPSIS analysis and life cycle thinking approach, respectively, so as to achieve a robustness and comprehensiveness of the remanufacturing design improvement;

e. Ensuring the operational, technological, economic and environmental considerations are well addressed and incorporated into the product recovery planning process through qualitative and quantitative analyses;

f. Utilizing the stochastic dynamic programming and probability theory to analyze the impact of the quality of the returned products on EOL decision making;

g. Adopting HALG to represent the product structure and the interconnections among components and thus enable the determination of the optimal depth of disassembly and the EOL fate for each component.

5.4   LIMITATIONS AND RECOMMENDATIONS

In this solutions guide, the proposed methodologies have been demonstrated with case studies to show their utility and applicability. However, there is a need for more case studies with different types of the products to further validate and improve the research findings and design tools. It should be noted that the solutions guide for remanufacturability assessment assumes that products are dealt with in a decoupled way, i.e., without considering the financial synergy resulted from sharing reverse logistic cost or setup cost with other product types within the factory. When a full-scale treatment of economic analysis is carried out, the methodology should be iterated with the readjusted cost value, such as the reverse logistic cost or resale price, in order to recalculate the EOL strategy on the single-product level. In addition, throughout this solution guide, energy has been used as the major environmental impact indicator, and the reason being that energy consumption has been confirmed by several studies (Sutherland et al., 2008; Huijbregts et al., 2006; Hula et al., 2003) as a major contributor to a number of environmental problems, e.g., global warming, acidification, eutrophication and stratospheric ozone depletion. If other factors of environmental impact are taken into account, like aquatic/terrestrial toxicity, smog formation, etc., the product EOL decision making might be different depending on weights assigned for different environmental impact categories.

There are a few issues that have not been considered in this solutions guide, which can be further explored and developed to improve the contributions made in this research.

a. Integration design for remanufacturing with PSS

The intense worldwide competition among manufacturers has motivated companies to shift the paradigm from product sales to service business models. This service business model is also referred to as "functional sales/economy", "product service combinations", "product-to-service" and "servicing and product service systems (PSS)". The reason for this shift is that companies have discovered the profit which could be gained during the product's use phase as well as the economic opportunities in the aftermarket of the product. An example of a service business model is when the companies provide the service of washing clothes instead of selling the actual washing machine. On the customer side, they only need to pay for the number of the laundry loads used instead of purchasing the washing machine (Sundin and Bras, 2005). This paradigm shift has led OEMs to focus more on product maintenance and remanufacturing (Sundin and Lindahl, 2008). This paradigm has also provided them with more incentive to improve the potential of the product remanufacturing through design, so as to extend the physical life cycle of their products and make profit from the PSS. This paradigm shift has called for insight and research work into product design requirement that would facilitate both service selling and remanufacturing and how this combination will work in practice (Hatcher et al., 2011).

b. Design for remanufacturing with an embedded sensor

Uncertainty in the quality and quantity of product return has been identified as one of the major issues that complicate the remanufacturing strategy planning process. To address these issues, embedding smart sensors in products has been proposed to monitor the product and obtain useful information, such as product identity, constituent components, remaining service life, remanufacturing history of the product to facilitate EOL decision making (Fang et al., 2013). An example is the use of RFID to retrieve, update and manage product information throughout the entire life cycle (Kiritsis et al., 2003; Parlikad and McFarlane, 2007). Despite the benefit of using embedded smart sensors for remanufacturing, there are still challenges and issues that limit the application of sensors, which should be addressed in the product design stage, such as the methods and location to mount the sensors without compromising the product functional performance and reliability, the capability of the sensors to store and transmit the information and the economic justification of installing embedded sensors. Further investigation on using embedded smart sensors is imperative to facilitate remanufacturing operations and decision making at the EOL stage.

c. Remanufacturing knowledge database

To assist decision makers in the application of the design for remanufacturing decision support tool and simplify the computation complexity, a computation tool based on Visual C# has been developed that allows for fast computation and ease of use of this methodology. The future work can focus on developing an "expert system" to automate the evaluation process to enable the users to input minimum rating for several criteria. This expert system can be achieved through gathering of remanufacturing expert knowledge, empirical evidence from case studies or theoretical derivation. Such expert systems would be of significance for users, especially designers who lack the required remanufacturing knowledge and understanding, when evaluating the design candidates with respect to remanufacturability.

Appendix

APPENDIX I: PRODUCT DESIGN GUIDELINES FOR REMANUFACTURING

TABLE I.1
Product Design Guidelines for Remanufacturing

Remanufacturing Process	Remanufacturing Requirement	Design Criteria
Reverse logistics	• Basic description of the product • Avoid damages during transportation	• Labels, graphical communication, packaging or even the form of a product could be positioned on the packaging • Sufficient clearance and support at the base • Avoid structures extruding outside
Disassembly	• Easy access to internal regions • Easy to loosen joints/fasteners • Reduce the variation of the tools used • Prevent part damage during the disassembly process • Prevent the corrosion of parts • Clear instruction of the product disassembly process • Easy access to the fastener/joints • Easy identification of the fastener • Using one disassembly direction • Multi-disassembly should be possible with one operation	• Time to remove items for access • Number of items to remove for access • Number of fasteners to remove • Number of different tools to unlock the joints • Number of permanent joints • Number of parts damaged • Number of fasteners damaged • Isolate the part from the elements • Use non-corrosive materials • Disassembly layout /instructions provided • Position of the parts • Type of fasteners/joints • Types of parts • Position of the fasteners/joints • Standardization of the fasteners/joints

(Continued)

TABLE I.1 (*Continued*)
Product Design Guidelines for Remanufacturing

Remanufacturing Process	Remanufacturing Requirement	Design Criteria
Sorting and inspection	• Ease of classification of the components • Ease of assessing the condition of the components • Request for more objective testing methods • Tools to facilitate the sorting process • Ease in detecting wear and corrosion • Component information are clearly indicated (life cycle, composition, wear indicator, etc.) • Testing points are easy to access	• Parts are identical or grossly dissimilar • Standardization of the parts • Small number of components and connections • Color coding/numbering system for similar parts • Small number of inspection tools • Simple part test • Descriptions of life cycle, composition, wear indicator are provided
Cleaning	• Accessibility of the internal parts • Simple method for cleaning • Simple inner and outside surfaces • Standard cleaning methods • Less wastes and health concerns • Less variation of the cleaning methods • Instruction for cleaning methods • Labels and instruction to withstand cleaning processes	• Number of cavities/corners difficult to clean • Surface roughness • Total waste generated • Time to clean • Total cleaning material used • Specify cleaning methods • Labels and instructions can withstand the cleaning process • Type of materials • Shape of the parts

(Continued)

TABLE I.1 (*Continued*)
Product Design Guidelines for Remanufacturing

Remanufacturing Process	Remanufacturing Requirement	Design Criteria
Reconditioning	• Parts are robust • Avoid subjective criteria • Fewer parts for replacements • Avoid technological or aesthetical obsolescence • Modularity updatable • Clear information of the product displayed • Texture areas are refurbishable	• Bulky – over design • Wear resistant surface design • Number of the usage cycles • Number of wear and failure prone positions • Number/cost of reparable components • Technological cycle of core components • Aesthetical cycle of core components • Component modularity • Upgradability of components • Contains a tracking method for life • Number of discarded components • Number of parts refurbished • Number of parts replaced
Reassembly and testing	• Ease for adjustments • Capable and adaptable for upgradability • Simple methods for testing	• Number of adjustments • Time to reassemble • Time of final testing • Upgraded configurations assembly without modification

APPENDIX II. MATHEMATICAL MODELS, SOFTWARE TOOLS OR STATICS REFERENCE FOR DESIGN FOR REMANUFACTURING

TABLE II.1
Mathematical Models, Software Tools or Statics Reference for Design for Remanufacturing (Partially Cited from Hatcher et al., 2011)

Approach	Author(s)	Format	Style	Key Purpose	Design Stage	Advantages	Disadvantages	Use in Industry
DfRem metrics	Bras and Hammond (1996); Amezquita et al. (1995)	Calculations/ software	Quant	Assess remanufacturability	Detail	• Process oriented • Familiar concept (DfMA)	• Complex • Retrospective • No guidance • Complex	No
DfRem tools	Yang et al. (2015)	Calculations/ software	Quant	Selection of most feasible design	Detail	• Lifecycle thinking	• Complex	No
RemPro matrix	Sundin (2004)	Reference	Qual	Guidance, prioritization of issues	Concept development	• Simple • Offers guidance process	• Subjective • No guidance	No
REPRO2	Zwolinski et al. (2006); Zwolinski and Brissaud (2008); Gehin et al. (2008)	Software	Qual	Decision making, provide past examples	Concept generation	• Early in design process • Does not require extensive knowledge • Offer guidance	• Subjective	No

(Continued)

TABLE II.1 (*Continued*)

Mathematical Models, Software Tools or Statics Reference for Design for Remanufacturing (Partially Cited from Hatcher et al., 2011)

Approach	Author(s)	Format	Style	Key Purpose	Design Stage	Advantages	Disadvantages	Use in Industry
DfRem guidelines	Ijomah (2009); Ijomah et al (2007);	Reference	Qual	Guidance	Concept generation	• Simple • Offers guidance	• Subjective • Lack lifecycle thinking	Unknown
DfRem Metric	Du et al. (2012)	Calculations/ software	Quant	Assess remanufacturability Suggest improvement	Detail/ redesign	• Offer guidance	• Complex • Retrospective	No
Hierarchical decision model	Lee et al. (2010)	Calculations	Quant	Design of product architecture for most profitable disassembly	Embodiment	• Lifecycle thinking	Not holistic	No
Energy comparison tool (CED)	Yang et al. (2014b)	Calculations	Quant	Compare product overall life cycle	Detail	• Lifecycle thinking	• No guidance	No
Component reliability assessment	Zhang et al. (2010)	calculation	Quant	Remanufacturing strategy decision making	Embodiment	• Customer focused • Process oriented	• Not holistic • No guidance	No

APPENDIX III: DESIGN AIDS THAT HAVE BEEN APPROPRIATED TO FACILITATE DFREM

TABLE III.1
Design Aids that Have Been Appropriated to Facilitate DfRem (Partially Cited from Hatcher et al., 2011)

Approach	Author(s)	Format	Style	Key Purpose	Design Stage	Advantage	Disadvantage	Use in Industry
Modularization	Wang et al. (2013)	Concept	Qual	Traditional: improve manufacturing efficiency Reman: ease of disassembly	Concept develops	• Familiar concept	• Not holistic • No guidance	Yes
FMEA	Abdullah et al. (2013)	Paper/software	Quant	Traditional: prioritize and prevent product failure Reman: reduce waster	Concept develops Redesign	• Familiar concept • Lifecycle thinking • Process oriented	• Not holistic • Reliant on reman-OEM feedback • No guidance	Yes
Platform design	King and Burgess (2005)	Concept	Quant	Traditional: reduce manufacturing costs and retain customer choice Reman: simplify process organization	Concept develops	• Familiar concept • Lifecycle thinking	• Not holistic • No guidance	Yes
Disassembly	Chiodo and Ijomah (2009)	Concept	Qual	Efficient disassembly	Concept develops	• Process oriented	• Not holistic	No
Design for Environment tools	Pigosso et al. (2009)	Various	Varies	Improve environmental performance	Various	• Lifecycle thinking	• Not holistic • complex	No
Fuzzy-QFD	Yang et al. (2013)	Paper/software	Quant/qual	Consider the voice of the remanufacturer, environment concern, economic consideration during early design stage	Concept develops	• Familiar concept • Process oriented	• Reliant on reman-OEM feedback	Yes

APPENDIX IV: SCREENSHOTS OF DFREM TOOL

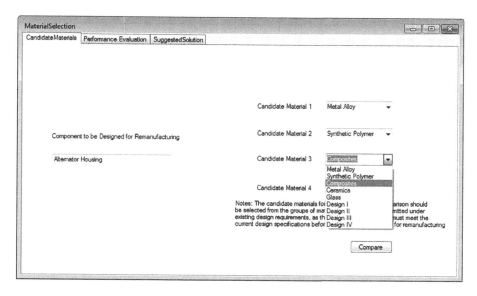

FIGURE IV.1   Define the component and material.

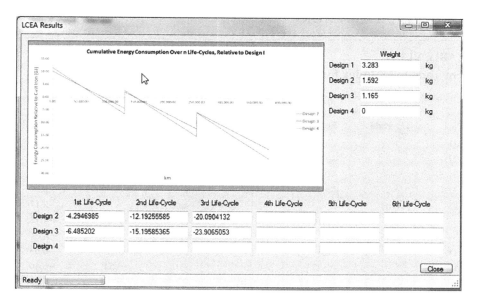

FIGURE IV.2   Calculate the environmental performance of candidate design.

APPENDIX V: ECONOMIC INDEX FOR DIFFERENT EOL OPTIONS (ALTERNATOR)

TABLE V.1
Economic Index for Different EOL Options (Alternator)

Part	q=1			q=2		
	Upgrade	Reman	Disposal	Upgrade	Reman	Disposal
21	-	−12.00	−35.00	-	−24.00	−35.00
22	-	−3.50	−6.00	-	−7.00	−6.00
31	-	−6.00	−2.00	-	−∞	−2.00
32	-	−4.80	−5.00	-	−∞	−5.00
25	−6.00	-	−8.00	−∞	-	−8.00
33	-	−5.50	−10.00	-	−11.00	−10.00
34	−8.00	-	−20.00	−16.00	-	−20.00
11	-	−4.00	−7.00	-	−8.00	−7.00
12	-	−6.00	−13.00	-	−12.00	−13.00
13	-	-	−41.00	-	-	−41.00
14	-	-	−50.00	-	-	−50.00
23	-	-	−7.00	-	-	−7.00
24	-	-	−30.00	-	-	−30.00

APPENDIX VI: ENVIRONMENTAL INDEX FOR DIFFERENT EOL OPTIONS (ALTERNATOR)

TABLE VI.1
Environmental Index for Different EOL Options (Alternator)

Part	q=1			q=2		
	Upgrade	Reman	Disposal	Upgrade	Reman	Disposal
21	-	−0.96	−12.06	-	−1.92	−12.06
22	-	−0.07	−2.42	-	−0.14	−2.42
31	-	−0.03	−0.71	-	−∞	−0.71
32	-	−0.03	−0.20	-	−∞	−0.20
25	−0.09	-	−2.17	−∞	-	−2.17
33	-	−0.16	−4.13	-	−0.32	−4.13
34	−0.96	-	−12.01	−1.93	-	−12.01
11	-	−0.11	−0.63	-	−0.22	−0.63
12	-	−0.27	−6.40	-	−0.54	−6.40
13	-	-	−12.48	-	-	−12.48
14	-	-	−20.22	-	-	−20.22
23	-	-	−0.91	-	-	−0.91
24	-	-	−16.14	-	-	−16.14

APPENDIX VII: CONDITIONAL PROBABILITY OF QUALITY LEVEL FOR SUBASSEMBLIES AND COMPONENTS (ALTERNATOR)

TABLE VII.1
Conditional Probability of Quality Level for Subassemblies and Components (Alternator)

L	Subassembly	Comp	Pr(1|1)	Pr(2|1)	Pr(1|2)	Pr(2|2)
1	0	11	0.8	0.2	0.4	0.6
		12	1	0	0.5	0.5
		13	0.5	0.5	0.2	0.8
		14	0.5	0.5	0.2	0.8
2	13	21	0.5	0.5	0.2	0.8
		22	0.5	0.5	0.2	0.8
	14	23	0.5	0.5	0.2	0.8
		24	0.8	0.2	0.5	0.5
		25	0.5	0.5	0.2	0.8
3	23	31	0.5	0.5	0.2	0.8
		32	0.5	0.5	0.2	0.8
	24	33	1	0	0.5	0.5
		34	1	0	0.4	0.6

APPENDIX VIII: DISASSEMBLY COST FOR DIFFERENT DISASSEMBLY STRATEGIES (ALTERNATOR)

TABLE VIII.1
Disassembly Cost for Different Disassembly Strategies (Alternator)

L	Subassembly	Break Joint	Independent Components	Connected Components	Cost
1	0	J11-12	[11]	[12,13,14]	0.4
		J13-14	[14]	[11,12,13]	1.8
		J12-13	[∅]	[11,12,13,14]	2.0
		J11-12, J12-13	[11][12]	[13,14]	2.4
		J12-13, J13-14	[13][14]	[11,12]	3.8
		J11-12,J12-13, J13-14	[11][12][13][14]	[∅]	4.2
		∅	[∅]	[11,12,13,14]	0.0

(Continued)

TABLE VIII.1 (*Continued*)
Disassembly Cost for Different Disassembly Strategies (Alternator)

L	Subassembly	Break Joint	Independent Components	Connected Components	Cost
2	13	J21-22	[21][22]	[⊘]	0.8
		⊘	[⊘]	[21,22]	0.0
	14	J23-25	[⊘]	[23,24,25]	1.0
		J23-24	[⊘]	[23,24,25]	1.0
		J24-25	[⊘]	[23,24,25]	1.0
		J23-25, J23-24	[23]	[24,25]	2.0
		J24-25, J23-24	[24]	[23,25]	2.0
		J23-25, J24-25	[25]	[23,24]	2.0
		J23-25, J23-24, J24-25	[23][24][25]	[⊘]	3.0
		⊘	[⊘]	[23,24,25]	0.0
3	23	J31-32	[31][32]	[⊘]	1.0
		⊘	[⊘]	[31,32]	0.0
	24	J33-34	[33][34]	[⊘]	1.0
		⊘	[⊘]	[33,34]	0.0

APPENDIX IX: ECONOMIC INDEX FOR DIFFERENT EOL OPTIONS (HEDGE TRIMMER)

TABLE IX.1

Economic Index for Different EOL Options (Hedge Trimmer)

Part		$q=1$			$q=2$	
	Upgrade	Reman	Disposal	Upgrade	Reman	Disposal
11		−0.81	−2.68		−2.44	−2.68
12		−0.41	−1.12		−∞	−1.12
33		−2.53	−8.30		−7.58	−8.30
34		−0.90	−3.00		−2.70	−3.00
23		−0.06	−0.20		−∞	−0.20
24		−0.81	−2.68		−2.44	−2.68
25		−0.18	−0.60		−∞	−0.60
41		−0.74	−2.44		−2.23	−2.44
42		−0.70	−2.29		−∞	−2.29
43		−1.59	−5.27		−4.77	−5.27
44		−0.24	−0.49		−∞	−0.49
45		−2.25	−6.26		−6.76	−6.26
46		−0.74	−2.41		−∞	−2.41
13		−∞	−30.45		−∞	−30.45
14		−∞	−3.47		−∞	−3.47
21		−∞	−19.16		−∞	−19.16
22		−∞	−11.29		−∞	−11.29
31		−∞	−4.73		−∞	−4.73
32		−∞	−14.43		−∞	−14.43

APPENDIX X: ENVIRONMENTAL INDEX FOR DIFFERENT EOL OPTIONS (HEDGE TRIMMER)

TABLE X.1

Environmental Index for Different EOL Options (Hedge Trimmer)

	$q=1$			$q=2$		
Part	Upgrade	Reman	Disposal	Upgrade	Reman	Disposal
11		−0.12	−3.67		−0.36	−3.67
12		−0.02	−0.75		−∞	−0.75
33		−0.23	−1.34		−0.69	−1.34
34		−0.01	−0.05		−0.03	−0.05
23		−0.01	−0.18		−∞	−0.18
24		−0.12	−3.67		−0.36	−3.67
25		−0.02	−0.52		−∞	−0.52
41		−0.07	−1.16		−0.20	−1.16
42		−0.07	−1.19		−∞	−1.19
43		−0.08	−1.37		−0.24	−1.37
44		−0.03	−1.18		−∞	−1.18
45		−0.08	−3.53		−0.24	−3.53
46		−0.10	−0.57		−0.30	−0.57
13		−∞	−10.40		−∞	−10.40
14		−∞	−4.37		−∞	−4.37
21		−∞	−9.01		−∞	−9.01
22		−∞	−1.39		−∞	−1.39
31		−∞	−2.35		−∞	−2.35
32		−∞	−6.66		−∞	−6.66

APPENDIX XI: CONDITIONAL PROBABILITY OF QUALITY LEVEL FOR SUBASSEMBLIES AND COMPONENTS (HEDGE TRIMMER)

TABLE XI.1

Conditional Probability of Quality Level for Subassemblies and Components (Hedge Trimmer)

L	Subassembly	Comp	Pr(1|1)	Pr(2|1)	Pr(1|2)	Pr(2|2)
1	0	11	0.8	0.2	0.2	0.8
		12	0.8	0.2	0.2	0.8
		13	0.8	0.2	0.2	0.8
		14	0.8	0.2	0.2	0.8
2	13	21	0.8	0.2	0.2	0.8
		22	0.8	0.2	0.2	0.8
	14	23	0.8	0.2	0.2	0.8
		24	0.8	0.2	0.2	0.8
		25	0.8	0.2	0.2	0.8
3	21	31	0.8	0.2	0.2	0.8
		32	0.8	0.2	0.2	0.8
	22	33	0.8	0.2	0.2	0.8
		34	0.8	0.2	0.2	0.8
4	31	41	0.8	0.2	0.2	0.8
		42	0.8	0.2	0.2	0.8
	32	43	0.8	0.2	0.2	0.8
		44	0.8	0.2	0.2	0.8
		45	0.8	0.2	0.2	0.8
		46	0.8	0.2	0.2	0.8

APPENDIX XII: DISASSEMBLY COST FOR DIFFERENT DISASSEMBLY STRATEGIES (HEDGE TRIMMER)

TABLE XII.1
Disassembly Cost for Different Disassembly Strategies (Hedge Trimmer)

L	Subassembly	Break Joint	Independent Components	Connected Components	Cost
1	0	J11-12	[11]	[12,13,14]	−2.7
		J11-12, J12-13	[11],[12]	[13,14]	−3.0
		J11-12, J12-13, J13-14	[11][12][13][14]	[∅]	−3.3
		∅	[∅]	[11,12,13,14]	0.0
	13	J21-22	[21],[22]	[∅]	−0.6
		∅	[∅]	[21,22]	0.0
	14	J23-24, J23-25	[23]	[24,25]	−0.8
		J23-24, J24-25	[24]	[23,25]	−0.8
		J23-25, J24-25	[25]	[23,24]	−0.8
		J23-24, J24-25, J23-25	[23],[24],[25]	[∅]	−1.2
		∅	[∅]	[23,24,25]	0.0
	21	J31-32	[31],[32]	[∅]	−1.5
		∅	[∅]	[31,32]	0.0
	22	J33-34	[33],[34]	[∅]	−2.5
		∅	[∅]	[33,34]	0.0
	31	J41-42	[41],[42]	[∅]	−0.9
		∅	[∅]	[41,42]	0.0
	32	J43-44	[44]	[43,45,46]	−1.5
		J45-46	[46]	[44,43,45]	−1.2
		J43-44, J44-45	[44],[43]	[45,46]	−2.1
		J45-46, J43-45	[45],[46]	[43,44]	−1.8
		J43-44, J45-46	[44],[46]	[43,45]	−2.7
		J43-44, J43-45, J45-46	[43],[44],[45],[46]	[∅]	−3.3

Bibliography

Abdullah T. A., Hashim A. and Baharudin A. B. (2013): Use of Failure Mode and Effects Analysis (FMEA) method in remanufacturing analysis for engine block, *Applied Mechanics and Materials*, Vol. 465, Iss. 10, pp. 1026–1033.

Activate (2015): Activate software for remanufacturing, retrieved 3rd Nov, 2015, from http://acctivate.com/features/manufacturing/remanufacturing/.

Adler D., Ludewig P., Kumar V. and Sutherland J. (2007): Comparing energy and other measures of environmental performance in the original manufacturing and remanufacturing of engine components, in *International Manufacturing Science and Engineering Conference*, American Society of Mechanical Engineers, pp. 851–860.

Amezquita T., Hammond R., Salazar M. and Bras B. (1995): Characterizing the remanufacturability of engineering systems, in *Proceedings of 1995 ASME Advances in Design Automation Conference*, Boston, MA, USA, pp. 271–278.

Andrea D. J. and Brown W. R. (1993): Material selection processes in the automotive industry, University of Michigan Transportation Research Institute, Report Number: UMTRI 93-40-5.

Anityasari M. and Kaebernick H. (2008): A concept of reliability evaluation for reuse and remanufacturing, *International Journal of Sustainable Manufacturing*, Vol. 1, Iss. 1/2, pp. 3–17.

Ashby M. F. (2012): *Materials and the Environment: Eco-Informed Material Choice*, Elsevier Science & Technology, Canada.

Atasu A., Guide Jr V. D. R. and Van Wassenhove L. N. (2010): So what if remanufacturing cannibalizes my new product sales, *California Management Review*, Vol. 52, Iss. 2, pp. 56–76.

Ayres R., Ferrer G. and Van Leynseele T. (1997): Eco-efficiency, asset recovery and remanufacturing, *European Management Journal*, Vol. 15, Iss. 5, pp. 557–574.

Bauccio M. (1993): *ASM Metals Reference Book*, ASM International.

Boran F. E., Genç S., Kurt M. and Akay D. (2009): A multi-criteria intuitionistic fuzzy group decision making for supplier selection with TOPSIS method, *Expert Systems with Applications*, Vol. 36, Iss. 8, pp. 11363–11368.

Boustead I. and Hancock G. F. (1979): *Handbook of Industrial Energy Analysis*, Ellis Horwood, Barcelona.

Bras B. and Hammond R. (1996): Towards design for remanufacturing—Metrics for assessing remanufacturability, in *1st International Workshop on Reuse*, Eindhoven, the Netherlands, pp. 35–52.

Bras B. and McIntosh M. W. (1999): Product, process, and organizational design for remanufacture–an overview of research, *Robotics and Computer-Integrated Manufacturing*, Vol. 15, Iss. 3, pp. 167–178.

Çalışkan H., Kurşuncu B., Kurbanoğlu C. and Güven Ş. Y. (2013): Material selection for the tool holder working under hard milling conditions using different multi criteria decision making methods, *Materials & Design*, Vol. 45, pp. 473–479.

Cao H. J., Liu F., Li C. B. and Liu C. (2006): An integrated method for product material selection considering environmental factors and a case study, *Materials Science Forum*, Vol. 532–533, pp. 1032–1035.

Cat® engines and engine systems provide power to the world, retrieved 26th Mar, 2015, from http://origin-www.cat.com/engines.

Chan J. W. (2008): Product End-of-Life options selection: Grey relational analysis approach, *International Journal of Production Research*, Vol. 46, Iss. 11, pp. 2889–2912.

Charter M. and Gray C. (2008): Remanufacturing and product design, *International Journal of Product Development*, Vol. 6, Iss. 3, pp. 375–392.

Chatterjee P. and Chakraborty S. (2012): Material selection using preferential ranking methods, *Materials & Design*, Vol. 35, pp. 384–393.

Chang M. M. L., Nee A. Y.C. and Ong S. K. (2020): Interactive AR-assisted product disassembly sequence planning (ARDIS), *International Journal of Production Research*, Vol. 58, Iss. 16, pp. 4916–4931.

Chang M. M. L., Ong S. K. and Nee A. Y. C. (2017a): Approaches and challenges in product disassembly planning for sustainability, *Procedia Cirp*, Vol. 60, pp. 506–511.

Chang M. M. L., Ong S. K. and Nee A. Y. C. (2017b): AR-guided product disassembly for maintenance and remanufacturing. *Procedia Cirp*, Vol. 61, pp. 299–304.

Chen C. T. (2000): Extensions of the TOPSIS for group decision-making under fuzzy environment, *Fuzzy Sets and Systems*, Vol. 114, Iss. 1, pp. 1–9.

Chen J. M. and Chang C. (2012): The economics of a closed-loop supply chain with remanufacturing, *Journal of the Operational Research Society*, Vol. 63, Iss. 10, pp. 1323–1335.

Chen L. H. and Hung C. C. (2010): An integrated fuzzy approach for the selection of outsourcing manufacturing partners in pharmaceutical R&D, *International Journal of Production Research*, Vol. 48, Iss. 24, pp. 7483–7506.

Chu C. H., Luh Y. P., Li T. C. and Chen H. (2009): Economical green product design based on simplified computer-aided product structure variation, *Computers in Industry*, Vol. 60, Iss. 7, pp. 485–500.

Corporation S. A. I. and Curran M. A. (2006): Life-cycle assessment: Principles and practice, National Risk Management Research Laboratory, Office of Research and Development, US Environmental Protection Agency.

Cristofari M., Deshmukh A. and Wang B. (1996): Green quality function deployment, in *Proceedings of the 4th International Conference on Environmentally Conscious Design and Manufacturing*, Cleveland, OH, USA, 23–25 July, pp. 297–304.

Curlee T. R., Das S., Rizy C. G. and Schexnayder S. M. (1994): Recent trends in automotive recycling: Energy and economic assessment, ORNL/TM-12628, Oak Ridge National Laboratory, Oak Ridge, TN.

David M. and Anderson, P.E. (2014): *Design for Manufacturability: How to Use Concurrent Engineering to Rapidly Develop Low-Cost, High-Quality Products for Lean Production*, CRC Press.

Davis J. R. (1996): *ASM Specialty Handbook: Cast Irons*, ASM International.

Dawson S. and Hang F. (2009): Compacted graphite iron-a material solution for modern diesel engine cylinder blocks and heads, *China Foundry*, Vol. 6, Iss. 3, pp. 241–246.

Dawson S. and Indra F. (2007): Compacted graphite iron-a new material for highly stressed cylinder blocks and cylinder heads, *Fortschritt Berichte-VDI Reihe 12 Verkehrstechnik Fahrzeugtechnik*, Vol. 639, Iss. 2, pp. 181–192.

Deb K., Pratap A., Agarwal S. and Meyarivan T. (2002): A fast and elitist multiobjective genetic algorithm: NSGA-II, Evolutionary Computation, *IEEE Transactions on Evolutionary Computation*, Vol. 6, Iss. 2, pp. 182–197.

Dehghan-Manshadi B., Mahmudi H., Abedian A. and Mahmudi R. (2007): A novel method for materials selection in mechanical design: Combination of non-linear normalization and a modified digital logic method, *Materials & Design*, Vol. 28, Iss. 1, pp. 8–15.

Dobrzański L. and Madejski J. (2006): Prototype of an expert system for selection of coatings for metals, *Journal of Materials Processing Technology*, Vol. 175, Iss. 1, pp. 163–172.

Dong T., Zhang L., Tong R. and Dong J. (2006): A hierarchical approach to disassembly sequence planning for mechanical product, *The International Journal of Advanced Manufacturing Technology*, Vol. 30, Iss. 5–6, pp. 507–520.

Du Y., Cao H., Liu F., Li C. and Chen X. (2012): An integrated method for evaluating the remanufacturability of used machine tool, *Journal of Cleaner Production*, Vol. 20, Iss. 1, pp. 82–91.

Durairaj S. K., Ong S. K. Nee, A. Y.C. and Tan R. B. (2002): Evaluation of life cycle cost analysis methodologies, *Corporate Environmental Strategy*, Vol. 9, Iss. 1, pp. 30–39.

Durairaj S. K., Ong S. K., Nee A. Y. C. and Tan R. B. (2003): A proposed tool to integrate environmental and economical assessments of products, *Environmental Impact Assessment Review*, Vol. 23, Iss. 1, pp. 51–72.

Durairaj S. K., Ong S. K., Tan R. B. and Nee A. Y. C. (2001): Environmental life cycle cost analysis of products. *Environmental Management and Health*, Vol. 12, Iss. 3, pp. 260–276.

Duque Ciceri N., Gutowski T. and Garetti M. (2010): A tool to estimate materials and manufacturing energy for a product, in *Proceedings of the 2010 IEEE International Symposium on Sustainable Systems & Technology (ISSST)*, Austin, TX, USA, pp. 1–6.

Dweiri F. and Al-Oqla F. M. (2006): Material selection using analytical hierarchy process, *International Journal of Computer Applications in Technology*, Vol. 26, Iss. 4, pp. 182–189.

Ertuğrul İ. and Karakaşoğlu N. (2008): Comparison of fuzzy AHP and fuzzy TOPSIS methods for facility location selection, *The International Journal of Advanced Manufacturing Technology*, Vol. 39, Iss. 7–8, pp. 783–795.

Fang H. C., Ong S. K. and Nee A. Y. C. (2013): Challenges and issues of using embedded smart sensors in products to facilitate remanufacturing, in *Proceedings of 20th CIRP International Conference on Life Cycle Engineering*, 17–19 April, Singapore, pp. 679–685.

Fang H. C., Ong S. K. and Nee A. Y. C. (2014): Product remanufacturability assessment based on design information, *Procedia CIRP*, Vol. 15, pp. 195–200.

Fang H. C., Ong S. K. and Nee A. Y. C. (2015): Product remanufacturability assessment and implementation based on design features, *Procedia CIRP*, Vol. 26, pp. 571–576.

Fang H. C., Ong S. K. and Nee A. Y. C. (2016): An integrated approach for product remanufacturing assessment and planning, *Procedia CIRP*, Vol. 40, pp. 262–267.

Fayazbakhsh K., Abedian A., Manshadi B. D. and Khabbaz R. S. (2009): Introducing a novel method for materials selection in mechanical design using Z-transformation in statistics for normalization of material properties, *Materials & Design*, Vol. 30, Iss. 10, pp. 4396–4404.

Feldmann K., Meedt O., Trautner S., Scheller H. and Hoffman W. (1999): The "green design advisor": A tool for design for environment, *Journal of Electronics Manufacturing*, Vol. 9, Iss. 01, pp. 17–28.

Ferguson M., Guide V. D., Koca E. and Souza G. C. (2009): The value of quality grading in remanufacturing, *Production and Operations Management*, Vol. 18, Iss. 3, pp. 300–314.

Florence and sustainable business phone, retrieved 26th Mar, 2015, from http://www.florenceinc.com/en/remanufacturing.php.

Gehin A., Zwolinski P. and Brissaud D. (2008): A tool to implement sustainable End-of-Life strategies in the product development phase, *Journal of Cleaner Production*, Vol. 16, Iss. 5, pp. 566–576.

Gerrard J. and Kandlikar M. (2007): Is European End-of-Life vehicle legislation living up to expectations? Assessing the impact of the ELV Directive on 'green' innovation and vehicle recovery, *Journal of Cleaner Production*, Vol. 15, Iss. 1, pp. 17–27.

Ghazalli Z. and Murata A. (2011): Development of an AHP–CBR evaluation system for remanufacturing: End-of-Life selection strategy, *International Journal of Sustainable Engineering*, Vol. 4, Iss. 1, pp. 2–15.

González B. and Adenso-Díaz B. (2005): A bill of materials-based approach for End-of-Life decision making in design for the environment, *International Journal of Production Research*, Vol. 43, Iss. 10, pp. 2071–2099.

Goodall P., Rosamond E. and Harding J. (2014): A review of the state of the art in tools and techniques used to evaluate remanufacturing feasibility, *Journal of Cleaner Production*, Vol. 81, pp. 1–15.

Govindan K., Khodaverdi R. and Jafarian A. (2013): A fuzzy multi criteria approach for measuring sustainability performance of a supplier based on triple bottom line approach, *Journal of Cleaner Production*, Vol. 47, pp. 345–354.

Grote C., Jones R., Blount G., Goodyer J. and Shayler M. (2007): An approach to the EuP Directive and the application of the economic eco-design for complex products, *International Journal of Production Research*, Vol. 45, Iss. 18–19, pp. 4099–4117.

Guesser W. L., Duran P. and Krause W. (2004): Compacted graphite iron-a new material for diesel engine cylinder blocks, retrieved 26th Mar, 2015, from http://www.tupy.com.br/downloads/guesser/compacted_graphite_iron_for_diesel.pdf.

Gungor A. and Gupta S. M. (1997): An evaluation methodology for disassembly processes, *Computers & Industrial Engineering*, Vol. 33, Iss. 1, pp. 329–332.

Hammond R., Amezquita T. and Bras B. (1998): Issues in the automotive parts remanufacturing industry: A discussion of results from surveys performed among remanufacturers, *Engineering Design and Automation*, Vol. 4, Iss. 1, pp. 27–46.

Hatcher G., Ijomah W. and Windmill J. (2011): Design for remanufacture: A literature review and future research needs, *Journal of Cleaner Production*, Vol. 19, Iss. 17, pp. 2004–2014.

Hertwich E. G., Pease W. S. and Koshland C. P. (1997): Evaluating the environmental impact of products and production processes: A comparison of six methods, *Science of the Total Environment*, Vol. 196, Iss. 1, pp. 13–29.

Huijbregts M. A., Rombouts L. J., Hellweg S., Frischknecht R., Hendriks A. J., van de Meent D., Ragas A. M., Reijnders L. and Struijs J. (2006): Is cumulative fossil energy demand a useful indicator for the environmental performance of products? *Environmental Science & Technology*, Vol. 40, Iss. 3, pp. 641–648.

Hula A., Jalali K., Hamza K., Skerlos S. J. and Saitou K. (2003): Multi-criteria decision-making for optimization of product disassembly under multiple situations, *Environmental Science & Technology*, Vol. 37, Iss. 23, pp. 5303–5313.

Hwang C. and Yoon K. (1981): *Multiple Attribute Decision Making: Methods and Applications, A State of the Art Survey*, Springer Verlag, Berlin.

Ijomah W., McMahon C., Hammond G. and Newman S. (2007): Development of robust design-for-remanufacturing guidelines to further the aims of sustainable development, *International Journal of Production Research*, Vol. 45, Iss. 18–19, pp. 4513–4536.

Ijomah W. L. (2009): Addressing decision making for remanufacturing operations and design-for-remanufacture, *International Journal of Sustainable Engineering*, Vol. 2, Iss. 2, pp. 91–102.

Ilgin M. A. and Gupta S. M. (2010): Environmentally conscious manufacturing and product recovery (ECMPRO): A review of the state of the art, *Journal of Environmental Management*, Vol. 91, Iss. 3, pp. 563–591.

Ilgin M. A. and Gupta S. M. (2011): Performance improvement potential of sensor embedded products in environmental supply chains, *Resources, Conservation and Recycling*, Vol. 55, Iss. 6, pp. 580–592.

Ipoint (2015): Sustainable products – The precondition for a circular economy, retrieve 3rd Nov, 2015, from http://www.ipoint-systems.com/.

Isaacs J., Gupta S. and Messac A. (1997): A decision tool to assess the impact of automobile design on disposal strategies, *Journal of Industrial Ecology*, Vol. 1, Iss. 4, pp. 19–33.

Jørgensen A. (2013): Social LCA—a way ahead? *The International Journal of Life Cycle Assessment*, Vol. 18, Iss. 2, pp. 296–299.

Jee D. H. and Kang K. J. (2000): A method for optimal material selection aided with decision making theory, *Materials & Design*, Vol. 21, Iss. 3, pp. 199–206.

Jiang Z., Zhang H. and Sutherland J. W. (2011): Development of multi-criteria decision making model for remanufacturing technology portfolio selection, *Journal of Cleaner Production*, Vol. 19, Iss. 17, pp. 1939–1945.

Jin X., Ni J. and Koren Y. (2011): Optimal control of reassembly with variable quality returns in a product remanufacturing system, *CIRP Annals-Manufacturing Technology*, Vol. 60, Iss. 1, pp. 25–28.

Johnson M. R. (2002): Evaluating remanufacturing and demanufacturing for extended producer responsibility and sustainable product management, PhD. Dissertations, University of Windsor.

Jun H.-B., Cusin M., Kiritsis D. and Xirouchakis P. (2007): A multi-objective evolutionary algorithm for EOL product recovery optimization: Turbocharger case study, *International Journal of Production Research*, Vol. 45, Iss. 18–19, pp. 4573–4594.

Kemna R., van Elburg M., Li W. and van Holsteijn R. (2005): Methodology Study Eco-design of Energy-using Products - MEEUP product cases report, VHK for European Commission, Netherlands.

Kerr W. and Ryan C. (2001): Eco-efficiency gains from remanufacturing: A case study of photocopier remanufacturing at Fuji Xerox Australia, *Journal of Cleaner Production*, Vol. 9, Iss. 1, pp. 75–81.

Kim H. J., Raichur V. and Skerlos S. J. (2008): Economic and environmental assessment of automotive remanufacturing: Alternator case study, in *Proceedings of the 2008 International Manufacturing Science and Engineering Conference*, Evanston, IL, USA, pp. 33–40.

Kin S. T. M., Ong S. K. and Nee A. Y. C. (2014): Remanufacturing process planning, *Procedia Cirp*, Vol. 15, pp. 189–194.

King A. and Barker S. (2007): Using the Delphi Technique to establish a robust research agenda for remanufacturing, in *Proceedings of 14th CIRP Conference on Life Cycle Engineering*, 11–13 June, Japan, pp. 219–224.

King A. and Burgess S. (2005): The development of a remanufacturing platform design: A strategic response to the Directive on Waste Electrical and Electronic Equipment, *Journal of Engineering Manufacture*, Vol. 219, Iss. 8, pp. 623–631.

Kiritsis D., Bufardi A. and Xirouchakis P. (2003): Research issues on product lifecycle management and information tracking using smart embedded systems, *Advanced Engineering Informatics*, Vol. 17, Iss. 3, pp. 189–202.

Klöpffer W. and Renner I. (2008): Life-cycle based sustainability assessment of products, *Environmental Management Accounting for Cleaner Production, Eco-Efficiency in Industry and Science*, Vol. 24, pp. 91–102.

Kobayashi H. (2005): Strategic evolution of eco-products: A product life cycle planning methodology, *Research in Engineering Design*, Vol. 16, Iss. 1, pp. 1–16.

Kobayashi H. (2006): A systematic approach to eco-innovative product design based on life cycle planning, *Advanced Engineering Informatics*, Vol. 20, Iss. 2, pp. 113–125.

Koffler C. and Rohde-Brandenburger K. (2010): On the calculation of fuel savings through lightweight design in automotive life cycle assessments, *The International Journal of Life Cycle Assessment*, Vol. 15, Iss. 1, pp. 128–135.

Krikke H., Van Harten A. and Schuur P. (1998): On a medium term product recovery and disposal strategy for durable assembly products, *International Journal of Production Research*, Vol. 36, Iss. 1, pp. 111–140.

Kuo T. C. (2006): Enhancing disassembly and recycling planning using life-cycle analysis, *Robotics and Computer-Integrated Manufacturing*, Vol. 22, Iss. 5, pp. 420–428.

Laforest V., Raymond G. and Piatyszek É. (2013): Choosing cleaner and safer production practices through a multi-criteria approach, *Journal of Cleaner Production*, Vol. 47, pp. 490–503.

Lee H. B., Cho N. W. and Hong Y. S. (2010): A hierarchical End-of-Life decision model for determining the economic levels of remanufacturing and disassembly under environmental regulations, *Journal of Cleaner Production*, Vol. 18, Iss. 13, pp. 1276–1283.

Levelseven (2015): Remanufacturing software from level seven, retrieved 3rd Nov, 2015, from http://www.lvlsvn.com/reman.

Liu Y., Ong S. K. and Nee A. Y. C. (2014): Modular design of machine tools to facilitate design for disassembly and remanufacturing, *Procedia CIRP*, Vol. 15, pp. 443–448.

Lund R. (1998): Remanufacturing: An American resource, in *Proceedings of the Fifth International Congress Environmentally Conscious Design and Manufacturing*, Rochester Institute of Technology, Rochester, NY, USA, pp. 1–6.

Lund R. T. and Mundial B. (1984): Remanufacturing: The experience of the United States and implications for developing countries, World Bank, Washington DC, Technical Paper, 31.

Lye S. W., Lee S. and Khoo M. (2002): ECoDE–An environmental component design evaluation tool, *Engineering with Computers*, Vol. 18, Iss. 1, pp. 14–23.

Mabee D. G., Bommer M. and Keat W. D. (1999): Design charts for remanufacturing assessment, *Journal of Manufacturing Systems*, Vol. 18, Iss. 5, pp. 358–366.

Mattioda R. A., Mazzi A., Canciglieri Jr. O. (2015): Determining the principal references of the social life cycle assessment of products, *The International Journal of Life Cycle Assessment*, Vol. 20, Iss. 8, pp. 1155–1165.

McGlothlin S. and Kroll E. (1995): Systematic estimation of disassembly difficulties: Application to computer monitors, in *Proceedings of the IEEE International Symposium on Electronics and the Environment*, Orland, FL, USA, pp. 83–88.

Mehta C. and Wang B. (2001): Green quality function deployment III: A methodology for developing environmentally conscious products, *Journal of Design and Manufacturing Automation*, Vol. 1, Iss. 1–2, pp. 1–16.

Milani A. and Shanian A. (2006): Gear material selection with uncertain and incomplete data. Material performance indices and decision aid model, *International Journal of Mechanics and Materials in Design*, Vol. 3, Iss. 3, pp. 209–222.

Mirhedayatian S. M., Vahdat S. E., Jelodar M. J. and Saen R. F. (2013): Welding process selection for repairing nodular cast iron engine block by integrated fuzzy data envelopment analysis and TOPSIS approaches, *Materials & Design*, Vol. 43, pp. 272–282.

Murray G. (1997): *Handbook of Materials Selection for Engineering Applications*, CRC Press.

Nasr N., Hilton B. and German R. (2011): A framework for sustainable production and a strategic approach to a key enabler: Remanufacturing, in *Proceedings of the 8th Global Conference on Sustainable Manufacturing*, 22–24 November, Abu Dhabi, pp. 191–196.

Nasr N. and Thurston M. (2006): Remanufacturing: A key enabler to sustainable product systems, in *Proceedings of LCE 2006*, Leuven, Belgium, pp. 15–18.

NCMS (2010): LCA/sustainability toolkit, national center for manufacturing sciences, retrieved 26th Mar, 2015, from http://www.ncms.org/wp-content/NCMS_files/sustainability/2012Sustainability/pmPresentations/1300_Mehta_NCMS_Sustainability_Tools_6_12_2012.pdf.

NCMS (2012): LCA/Sustainability Toolkit, 2010. National Center for Manufacturing Sciences, retrieved 12th Jan, 2015, from http://www.ncms.org/wp-content/NCMS_files/sustainability/2012Sustainability/pmPresentations/1300_Mehta_NCMS_Sustainability_Tools_6_12_2012.pdf.

NCMS (2015): Remanufacturing assessment tool, retrieved 3rd Nov, 2015, from http://remanassesmenttool.reman.rit.edu/default.aspx?~w-ncrrr/RemanAssessmentTool/.

Opricovic S. (2011): Fuzzy VIKOR with an application to water resources planning, *Expert Systems with Applications*, Vol. 38, Iss. 10, pp. 12983–12990.

Opricovic S. and Tzeng G. H. (2004): Compromise solution by MCDM methods: A comparative analysis of VIKOR and TOPSIS, *European Journal of Operational Research*, Vol. 156, Iss. 2, pp. 445–455.

Ong S. K., Chang M. M. L. and Nee A. Y. C. (2021): Product disassembly sequence planning: State-of-the-art, challenges, opportunities and future directions, *International Journal of Production Research*, Vol. 59, Iss. 11, pp. 3493–3508.

Parlikad A. K. and McFarlane D. (2007): RFID-based product information in End-of-Life decision making, *Control Engineering Practice*, Vol. 15, Iss. 11, pp. 1348–1363.

Pbioore K., Gungor A. and Gupta S. M. (1998): Disassembly process planning using Petri nets, in *Proceedings of the 1998 IEEE International Symposium on Electronics and the Environment*, Oak Brook, IL, USA, pp. 88–93.

Peng A.-H. and Xiao X.-M. (2013): Material selection using PROMETHEE combined with analytic network process under hybrid environment, *Materials & Design*, Vol. 47, pp. 643–652.

Quella F. and Schmidt W.-P. (2003): Integrating environmental aspects into product design and development the new ISO TR 14062, *The International Journal of Life Cycle Assessment*, Vol. 8, Iss. 2, pp. 113–114.

Quigley F., Buggy M. and Birkinshaw C. (2002): Selection of elastomeric materials for compliant-layered total hip arthroplasty, *Proceedings of the Institution of Mechanical Engineers, Part H: Journal of Engineering in Medicine*, Vol. 216, Iss. 1, pp. 77–83.

Rajendran R. (2012): Gas turbine coatings–An overview, *Engineering Failure Analysis*, Vol. 26, pp. 355–369.

Rao R. V. (2008): A decision making methodology for material selection using an improved compromise ranking method, *Materials & Design*, Vol. 29, Iss. 10, pp. 1949–1954.

Rathod M. K. and Kanzaria H. V. (2011): A methodological concept for phase change material selection based on multiple criteria decision analysis with and without fuzzy environment, *Materials & Design*, Vol. 32, Iss. 6, pp. 3578–3585.

Rebitzer G., Ekvall T., Frischknecht R., Hunkeler D., Norris G., Rydberg T., Schmidt W.-P., Suh S., Weidema B. P. and Pennington D. (2004): Life cycle assessment: Part 1: Framework, goal and scope definition, inventory analysis, and applications, *Environment International*, Vol. 30, Iss. 5, pp. 701–720.

Remery M., Mascle C. and Agard B. (2012): A new method for evaluating the best product End-of-Life strategy during the early design phase, *Journal of Engineering Design*, Vol. 23, Iss. 6, pp. 419–441.

Robèrt K.-H., Schmidt-Bleek B., Aloisi de Larderel J., Basile G., Jansen J. L., Kuehr R., Price Thomas P., Suzuki M., Hawken P. and Wackernagel M. (2002): Strategic sustainable development—selection, design and synergies of applied tools, *Journal of Cleaner Production*, Vol. 10, Iss. 3, pp. 197–214.

Rose C. M. (2000): Design for environment: A method for formulating product End-of-Life strategies, PhD dissertation, Department of mechanical engineering, Stanford University.

Sahni S., Boustani A., Gutowski T. and Graves S. (2010): Engine remanufacturing and energy savings, Environmentally Benign Laboratory, Laboratory for Manufacturing and Productivity, Sloan School of Management, MITEI-1-d-2010.

Sakao T. (2007): A QFD-centred design methodology for environmentally conscious product design, *International Journal of Production Research*, Vol. 45, Iss. 18–19, pp. 4143–4162.

Samvedi A., Jain V. and Chan F. T. (2013): Quantifying risks in a supply chain through integration of fuzzy AHP and fuzzy TOPSIS, *International Journal of Production Research*, Vol. 51, Iss. 8, pp. 2433–2442.

Schau E. M., Traverso M. and Finkbeiner M. (2012): Life cycle approach to sustainability assessment: A case study of remanufactured alternators, *Journal of Remanufacturing*, Vol. 2, Iss. 1, pp. 1–14.

Shanian A., Milani A., Carson C. and Abeyaratne R. (2008): A new application of ELECTRE III and revised Simos' procedure for group material selection under weighting uncertainty, *Knowledge-Based Systems*, Vol. 21, Iss. 7, pp. 709–720.

Shanian A. and Savadogo O. (2006): TOPSIS multiple-criteria decision support analysis for material selection of metallic bipolar plates for polymer electrolyte fuel cell, *Journal of Power Sources*, Vol. 159, Iss. 2, pp. 1095–1104.

Shannon C. E. and Weaver W. (1947): *The Mechanical Theory of Communication*, University of Illinois Press.

Shemshadi A., Shirazi H., Toreihi M. and Tarokh M. J. (2011): A fuzzy VIKOR method for supplier selection based on entropy measure for objective weighting, *Expert Systems with Applications*, Vol. 38, Iss. 10, pp. 12160–12167.

Shu L. H. and Flowers W. C. (1999): Application of a design-for-remanufacture framework to the selection of product life-cycle fastening and joining methods, *Robotics and Computer-Integrated Manufacturing*, Vol. 15, Iss. 3, pp. 179–190.

Siew C. Y., Chang M. M. L., Ong S. K. and Nee A. Y. C. (2020): Human-oriented maintenance and disassembly in sustainable manufacturing, *Computers & Industrial Engineering*, Vol. 150, p. 106903.

Siew C. Y., Ong S. K. and Nee A. Y. C. (2021): Improving maintenance efficiency and safety through a human-centric approach, *Advances in Manufacturing*, Vol. 9, Iss. 1, pp. 104–114.

Smil V. (2008): *Energy in Nature and Society: General Energetics of Complex Systems*, MIT press.

Smith V. M. and Keolelan G. A. (2004): The value of remanufactured engines: Life-cycle environmental and economic perspectives, *Journal of Industrial Ecology*, Vol. 8, Iss. 1–2, pp. 193–221.

Soh S. L., Ong S. K. and Nee A. Y. C. (2014): Design for disassembly for remanufacturing: Methodology and technology, *Procedia CIRP*, Vol. 15, pp. 407–412.

Soh S. L., Ong S. K. and Nee A. Y. C. (2016): Design for assembly and disassembly for remanufacturing. Assembly Automation.

Song C., Guan X. and Zhao Q. (2005): Machine learning approach for determining feasible plans of a remanufacturing system, *IEEE Transactions on Automation Science and Engineering*, Vol. 3, Iss. 2, pp. 262–275.

Sprow E. (1992): The mechanics of remanufacture, *Manufacturing Engineering*, Vol. 108, Iss. 3, pp. 38–45.

Staikos T. and Rahimifard S. (2007): A decision-making model for waste management in the footwear industry, *International Journal of Production Research*, Vol. 45, Iss. 18–19, pp. 4403–4422.

Steinhilper R. (1998): Remanufacturing: The ultimate form of recycling, *Journal of Industrial Ecology*, Vol. 3, Iss. 2–3, pp. 189–192.

Subramoniam R., Huisingh D. and Chinnam R. B. (2009): Remanufacturing for the automotive aftermarket-strategic factors: Literature review and future research needs, *Journal of Cleaner Production*, Vol. 17, Iss. 13, pp. 1163–1174.

Subramoniam R., Huisingh D. and Chinnam R. B. (2010): Aftermarket remanufacturing strategic planning decision-making framework: Theory & practice, *Journal of Cleaner Production*, Vol. 18, Iss. 16, pp. 1575–1586.

Sullivan J. L. and Hu J. (1995): Life cycle energy analysis for automobiles, SAE Technical Paper 970663, Society of Automotive Engineers, DOI:10.4271/951829.

Sundin E. (2004): Product and process design for successful remanufacturing, Linköping Studies in Science and Technology Dissertation, No. 906.

Sundin E. and Bras B. (2005): Making functional sales environmentally and economically beneficial through product remanufacturing, *Journal of Cleaner Production*, Vol. 13, Iss. 9, pp. 913–925.

Sundin E. and Lindahl M. (2008): Rethinking product design for remanufacturing to facilitate integrated product service offerings, in *Proceedings of IEEE International Symposium on Electronics and the Environment*, San Francisco, CA, USA, pp. 1–6.

Sutherland J., Jenkins T. and Haapala K. (2010): Development of a cost model and its application in determining optimal size of a diesel engine remanufacturing facility, *CIRP Annals-Manufacturing Technology*, Vol. 59, Iss. 1, pp. 49–52.

Sutherland J. W., Adler D. P., Haapala K. R. and Kumar V. (2008): A comparison of manufacturing and remanufacturing energy intensities with application to diesel engine production, *CIRP Annals-Manufacturing Technology*, Vol. 57, Iss. 1, pp. 5–8.

The WEEE directive, retrieved 26th Mar, 2015, from http://www.recyclingyourmobile.co.uk/weee-directive.htm.

Toktay L. B. and Wei D. (2011): Cost allocation in manufacturing–remanufacturing operations, *Production and Operations Management*, Vol. 20, Iss. 6, pp. 841–847.

Umeda Y., Takata S. and Kimura F. (2012): Toward integrated product and process life cycle planning—An environmental perspective, *CIRP Annals-Manufacturing Technology*, Vol. 61, Iss. 2, pp. 681–702.

Vartabedian R. (1992): Iron or aluminum engines? Debate continues, *Los Angeles Times*, Collections.

Vatsayan, U., Pandey, K. M. and Biswas, A. (2014): Effects of heat treatment on materials used in automobiles: A case study, *Journal of Mechanical and Civil Engineering*, Vol. 11, Iss. 5, pp. 90–95.

Veerakamolmal P. and Gupta S. M. (1999): A combinatorial cost-benefit analysis methodology for designing modular electronic products for the environment, in *Proceedings of the 1999 IEEE International Symposium*, Boston, MA, USA, pp. 268–273.

Wang P. J., Liu Y., Ong S. K. and Nee A. Y. C. (2014): Modular design of machine tools to facilitate design for disassembly and remanufacturing, *Procedia CIRP*, Vol, 15, pp. 443–448.

Wang X. and Chan H. K. (2013a): A hierarchical fuzzy TOPSIS approach to assess improvement areas when implementing green supply chain initiatives, *International Journal of Production Research*, Vol. 51, Iss. 10, pp. 3117–3130.

Wang X. and Chan H. K. (2013b): An integrated fuzzy approach for evaluating remanufacturing alternatives of a product design, *Journal of Remanufacturing*, Vol. 3, Iss. 1, pp. 1–19.

Xu Y. and Feng W. (2014): Develop a cost model to evaluate the economic benefit of remanufacturing based on specific technique, *Journal of Remanufacturing*, Vol. 4, Iss. 1, pp. 4–12.

Yang S., MR A. R., Kaminski J. and Pepin, H. (2018): Opportunities for industry 4.0 to support remanufacturing, *Applied Sciences*, Vol. 8, Iss. 7, p. 1177.

Yang S., Ong S. K. and Nee A. Y. C. (2013): Design for remanufacturing-a Fuzzy-QFD approach. In *Re-engineering Manufacturing for Sustainability* (pp. 655–661). Springer, Singapore.

Yang S. S., Nasr N., Ong S. K. and Nee A. Y. C. (2016): A holistic decision support tool for remanufacturing: End-of-Life (EOL) strategy planning, *Advances in Manufacturing*, Vol. 4, Iss. 3, pp. 189–201.

Yang S. S., Nasr N., Ong S. K. and Nee A. Y. C. (2017): Designing automotive products for remanufacturing from material selection perspective, *Journal of Cleaner Production*, Vol. 153, pp. 570–579.

Yang S. S., Ngiam H. Y., Ong S. K. and Nee A. Y. C. (2015a): The impact of automotive product remanufacturing on environmental performance, *Procedia CIRP*, Vol. 29, pp. 774–779.

Yang S. S., Ong S. K. and Nee A. Y. C. (2015b): EOL strategy planning for components of returned products, *The International Journal of Advanced Manufacturing Technology*, Vol. 77, Iss. 5, pp. 991–1003.

Yang S. S., Ong S. K. and Nee A. Y. C. (2015c): Towards implementation of DfRem into the product development process, *Procedia CIRP*, Vol. 26, pp. 565–570.

Yang S. S., Ong S. K. and Nee A. Y. C. (2016): A decision support tool for product design for remanufacturing, *Procedia Cirp*, Vol. 40, pp. 144–149.

Yüksel H. (2010): Design of automobile engines for remanufacture with quality function deployment, *International Journal of Sustainable Engineering*, Vol. 3, Iss. 3, pp. 170–180.

Yurdakul M. and Ic Y. T. (2009): Application of correlation test to criteria selection for multi criteria decision making (MCDM) models, *The International Journal of Advanced Manufacturing Technology*, Vol. 40, Iss. 3–4, pp. 403–412.

Zeydan M. and Çolpan C. (2009): A new decision support system for performance measurement using combined fuzzy TOPSIS/DEA approach, *International Journal of Production Research*, Vol. 47, Iss. 15, pp. 4327–4349.

Zhang X., Zhang H., Jiang Z. and Wang Y. (2013): A decision-making approach for End-of-Life strategies selection of used parts, *The International Journal of Advanced Manufacturing Technology*, DOI: 10.1007/s00170-013-5234-0.

Zhang Y. (1999): Green QFD-II: A life cycle approach for environmentally conscious manufacturing by integrating LCA and LCC into QFD matrices, *International Journal of Production Research*, Vol. 37, Iss. 5, pp. 1075–1091.

Ziout A., Azab A. and Atwan M. (2014): A holistic approach for decision on selection of End-of-Life products recovery options, *Journal of Cleaner Production*, Vol. 65, pp. 497–516.

Zwolinski P., Lopez-Ontiveros M. A. and Brissaud D. (2006): Integrated design of remanufacturable products based on product profiles, *Journal of Cleaner Production*, Vol. 14, Iss. 15, pp. 1333–1345.

Index

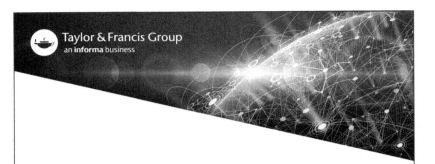

Printed in the United States
by Baker & Taylor Publisher Services